WELFARE AND HUSBANDRY OF CALVES

CURRENT TOPICS IN VETERINARY MEDICINE AND ANIMAL SCIENCE

VOLUME 19

Other titles in this series

Volume 1
Control of Reproduction in the Cow
edited by J.M. Sreenan

Volume 2
Patterns of Growth and Development in Cattle
edited by H. de Boer and J. Martin

Volume 3
Respiratory Diseases in Cattle
edited by W.B. Martin

Volume 4
Calving Problems and Early Viability of the Calf
edited by B. Hoffmann, I.L. Mason and J. Schmidt

Volume 5
The Future of Beef Production in the European Community
edited by J.C. Bowman and P. Susmel

Volume 6
Diseases of Cattle in the Tropics: Economic and Zoonotic Relevance
edited by M. Ristic

Volume 7
Control of Reproductive Functions in Domestic Animals
edited by W. Jöchle and D.R. Lamond

Volume 8
The Laying Hen and its Environment
edited by R. Moss

Volume 9
Epidemiology and Control of Nematodiasis in Cattle
edited by P. Nansen, R.J. Jørgensen and E.J.L. Soulsby

Volume 10
The Problem of Dark-Cutting in Beef
edited by D.E. Hood and P.V. Tarrant

Volume 11
The Welfare of Pigs
edited by W. Sybesma

Volume 12
The Mucosal Immune System
edited by F.J. Bourne

Volume 13
Laboratory Diagnosis in Neontal Calf and Pig Diarrhoea
edited by P.W. de Leeuw and P.A.M. Guinée

Volume 14
Advances in the Control of Theileriosis
edited by A.D. Irvin, M.P. Cunningham and A.S. Young

Volume 15
Fourth International Symposium on Bovine Leukosis
edited by O.C. Straub

Volume 16
Muscle Hypertrophy of Genetic Origin and its Use to Improve Beef Production
edited by J.W.B. King and F. Ménissier

Volume 17
Aujeszky's Disease
edited by G. Wittman and S.A. Hall

Volume 18
Transport of Animals Intended for Breeding, Production and Slaughter
edited by R. Moss

Series ISBN: 90-247-2429-5

WELFARE AND HUSBANDRY OF CALVES

A Seminar in the CEC Programme of Coordination of Research on Animal Welfare, organised by J.P. Signoret, and held in Brussels, 9-10 July, 1981

Sponsored by the Commission of the European Communities, Directorate-General for Agriculture, Coordination of Agricultural Research

Edited by

J.P. Signoret
I.N.R.A.
Centre de Recherches de Tours,
Nouzilly,
France

1982
MARTINUS NIJHOFF PUBLISHERS
THE HAGUE / BOSTON / LONDON
for
THE COMMISSION OF THE EUROPEAN COMMUNITIES

Distributors

for the United States and Canada
Kluwer Boston, Inc.
190 Old Derby Street
Hingham, MA 02043
USA

for all other countries
Kluwer Academic Publishers Group
Distribution Center
P.O. Box 322
3300 AH Dordrecht
The Netherlands

Library of Congress Cataloging in Publication Data
Main entry under title:

Welfare and husbandry of calves.

(Current topics in veterinary medicine and animal science ; v. 19)
1. Calves--Congresses. I. Signoret, J. P. II. Commission of the European Communities. Directorate-General for Agriculture. III. Series.
SF205.W44 1982 636.2'01 82-7875
ISBN 90-247-2680-8 AACR2

ISBN 90-247-2680-8 (this volume)
ISBN 90-247-2429-5 (series)

Publication arranged by
Commission of the European Communities,
Directorate-General Information Market and Innovation

EUR 7914 EN

Manuscript Preparation by
Janssen Services, 33a High Street, Chislehurst, Kent BR7 5AE, UK

PRINTED IN THE NETHERLANDS

CONTENTS

PREFACE

The Commission of the European Communities (CEC), Directorate General for Agriculture, Brussels, as part of the work of the Division Coordinating Agricultural Research, 9 - 10 July 1981, has continued a series of scientific meetings designed to review the current knowledge of the welfare of domestic animals with this seminar devoted to the calf.

Cattle make up a most important section of the animal industry in the European Communities (EC) and, with the exception of beef cattle, most calves have to undergo the disruption that exists in a so-called normal life. They are fed with milk replacers, segregated from adults, and so on. The feeding, as a non ruminant, of a future ruminant is not easy for the nutritionist. The pathologist faces difficulties in caring for the very young animal. The ethologist is concerned with the elimination of maternal care and the behaviour of the calf from birth. Furthermore public opinion, as a result of the great emotional reactions to the newly born, shows a much greater concern for the living conditions of the young than the adult farm animal.

This scientific seminar, orientated to welfare considerations, has brought together a range of opinions from the Member States of the EC where calf rearing is far from being homogeneous.

Veal production is important in France and Italy. Beef cattle breeds are mostly British and French. The Friesian and Holstein breeds exert a major impact in milk production. Despite the different eventual goal for the calf, viz. dairy cow, beef, ox, breeding bull or a veal calf, every breeder of cattle is very concerned with calf rearing. Whilst the scientist is more preoccupied by estimates of requirements in respect of physiology, nutrition, health and behaviour thereafter any objective evaluation of the various systems in practice from a welfare point of view must examine the level of production, return to the farmer, and the price to the consumer of the end-product.

The CEC thanks the participants for the papers presented, the organiser and Janssen Services for preparation of the manuscript of the Proceedings. The meeting has been a useful step to preparing more efficient and acceptable methods of cattle husbandry.

J.P. Signoret

SESSION I

BEHAVIOURAL AND PHYSIOLOGICAL REQUIREMENTS OF CALVES

Chairman: R. Moss

MOTHER - CALF BEHAVIOUR DURING THE FIRST SIX HOURS AFTER PARTURITION

Else Hermann and Nanny Stenum

Zoological Laboratory, University of Copenhagen,
Universitetspaiken 15, DK 2100, Copenhagen Ø, Denmark.

ABSTRACT

The behaviour of 18 mother-young pairs from two different stocks, one of cross-bred beef and dairy cattle and one of purebred dairy cattle, was observed during the first 6 h post partum. Particular interest was paid to the following in dams: isolation seeking, position, licking and vocalisation, maternal orientation, maternal rejection, and placentophagy. It was concluded that the primiparous dams (dairy heifers) were less effective as mothers compared to the multiparous dams, especially when compared to the cross-bred beef and dairy cows. Behaviour observed in calves was behaviour from birth to the initial phase of teat-seeking, teat-seeking and suckling activities. The activities of the dairy heifers' calves were generally delayed, while the dairy cows' calves deviated, especially at the time of first suckling.

INTRODUCTION

In Denmark calf mortality is relatively high, i.e. 15 - 16%. One of the factors involved, we believe, is the fact that cow and calf, as a rule, are separated immediately after the birth of the calf. The calf is thereby denied the opportunity of getting the essential colostrum by suckling naturally. This may influence calf welfare, because a calf is born practically without immunoglobulins.

Investigations (Smith et al., 1967; Selman et al., 1970; Selman et al., 1971 - all cited in Selman et al., 1971) have shown that calves suckling dams instead of being bucket fed had the highest serum immunoglobulin concentrations and therefore a higher chance of survival.

Other aspects of calf welfare may also be affected by separating calf and cow. Consequently it must be of great importance to analyse the general maternal-neonatal activities.

Maternal-filial behaviour in domesticated ungulates is well documented for sheep and goats. The period immediately following parturition has been observed by several workers (Alexander and Williams, 1964 and 1966; Bareham, 1976; Blauvelt, 1954; Collias, 1956; Hersher et al., 1963; Smith, 1965; Stephens and Linzell, 1974).

Studies on the maternal-filial behaviour of cattle seem to be more limited, especially when one focuses attention on the period immediately following birth. Some authors touch aspects of the subject (Arthur, 1961; Brownlee, 1950; Edwards and Broom, 1976; Finger and Brummer, 1969; Fraser, 1968; George and Barger, 1974; Hafez, 1964; Hafez and Bouissou, 1974; Kiley, 1973; McBride et al., 1967; Sambraus, 1971 and 1973; Schloeth, 1958; Stenbæk et al., 1980), but only a few have focussed attention on the development of natural mother-young relationships immediately upon birth (Donaldson, 1970; Hafez and Lineweaver, 1968; Selmann et al., 1970; Walker, 1950).

This investigation was made with the purpose of gaining more information about cow and calf activities during the first 6 h *post partum*. In addition, a comparison between three observed groups was made. It must be stated that differences in behaviour between the three groups may have been caused by several factors, among which are differences in breed, in rearing conditions, environment, number of calvings, and difficulties at labour. We have attempted to relate cause and effect in those cases where it was found reasonable to do so.

MATERIALS AND METHODS

The observations involved 18 mother-young pairs from two different herds, one of crossbred beef and dairy cattle and one of purebred dairy cattle. In the beef-dairy herd 8 mother-young pairs were observed of which all dams were multiparous (BD-cows). In the dairy herd (Holstein-Friesian race) dams were multiparous in 6 pairs (D-cows), and primiparous in 4 pairs (D-heifers)

The BD-cows and calves were part of a research herd kept in Mols Bjerge (Fussingø State Forestry District) and did not enter into any commercial farming production. During the winter the animals were kept in a cattle yard with free access to an outdoor pen, in the spring they were allowed to move freely between the cattle yard and a paddock, while the summer was spent on the pasture without access to the cattle yard. The calves were not separated at all from the cows during their first summer. The observations were carried out during the spring and early summer of 1979. One calving was observed in the cattle yard, two in the paddock, and five on the pasture.

The D-cows/heifers and calves were part of a normal dairy farm herd belonging to Mr. Bent Vithen in Lading (north west of Aarhus), where the animals were kept in a parlour and yard system. At the expected time of parturition a pregnant animal was transferred to a special calving box, measuring 3.10 x 3.00 m and covered with deep straw. The time from the transfer to the actual parturition was normally between 24 and 48 h. After parturition the mother-young pair was left together for 72 h, after which the cow was returned to the cattle yard and the calf transferred to another box. The observations were carried out during the winters of 1978 - 79 and 1979 - 80. Fifteen of the dam-calf pairs were observed for 6 h while 3 pairs, 1 BD-cow and 2 D-cows, were observed for 3 h *post partum*. Assistance at parturition was necessary in 6 cases, 2 D-cows and 4 D-heifers, while 12 dams were undisturbed. Neither the presence of the observer nor the use of light at night appeared to disturb the animals.

The observations were made continuously for 6 h *post partum*, starting at the moment the calf was born, i.e. completely free of the mother (time = 0). A stopwatch was used for recording the time, and the behaviour of the calf and the dam respectively was noted within every minute.

Listed below are the essential behavioural elements recorded:

The Dam

1. Position of the dam. 2. Licking of the calf (and, if possible, which parts of the body). 3. Cleaning of the birth area, including the eating of placenta. 4. Vocalisation. 5. Intraspecific agonistic behaviour. 6. Other activities: eating, rumination, defaecation and urination.

The Calf

1. Attempts to stand, i.e. the calf putting at least one foot to the ground, with some part of the body raised from the ground. 2. Standing position, including a) upright on all four legs for the first time, and b) standing for the first time, i.e. remaining upright for at least 5 min,

without falling over. 3. Teat-seeking, i.e. the calf pushing the muzzle against and possibly sucking the dam's body (any other specific associate or non-animal objects). 4. Suckling, i.e. the calf having a teat in the mouth and apparently obtaining milk. 5. Vocalisation. 6. Other activities: butting with the head, tailwagging, defaecation and urination.

It must be emphasised that the recording of the observed activities was a recording of whether or not a certain activity appeared within a certain minute.

The Mann-Whitney U-test was used for statistical hypothesis testing, the Spearman's Rho in tests for correlation and for trend, and the Chi-Square Test for differences in probabilities.

In drawing inferences from the tests the level of significance has been set to p 0.050, while the term 'a pronounced tendency' is used for p-values just above 0.050 and the term 'a tendency' is used for p-values up to 0.100.

MATERNAL BEHAVIOUR DURING THE FIRST SIX HOURS AFTER PARTURITION

The recorded activities were as follows, starting with the most homogeneous group, the BD-cows.

With the exception of the cow calving in the cattle yard, the BD-cows were seldom seen in the main body of the herd when parturition was near. They stayed at the periphery of, or at a distance from, the herd, often near the fenceline. When the main herd moved, the cow remained at the same spot and did not follow.

Six out of eight cows completed parturition in a lying position. The cow lay on one side with all four legs parallel to the ground and her head resting on the ground. The remaining two cows rose to their feet when the calf was born to its hips and then dropped the calf.

The six cows lying down rose to their feet, turned around, and started licking the calf less than one minute after the calf had been born, as did the two cows rising when the calf was born to its hips.

The cow's licking of the calf was accompanied by a quiet grunting sound. This was the dominant behaviour of the cow during the first hour after parturition. Thereafter licking the calf and vocalising, as well as the time spent in the standing position decreased during the first six hours after parturition (Figure 1).

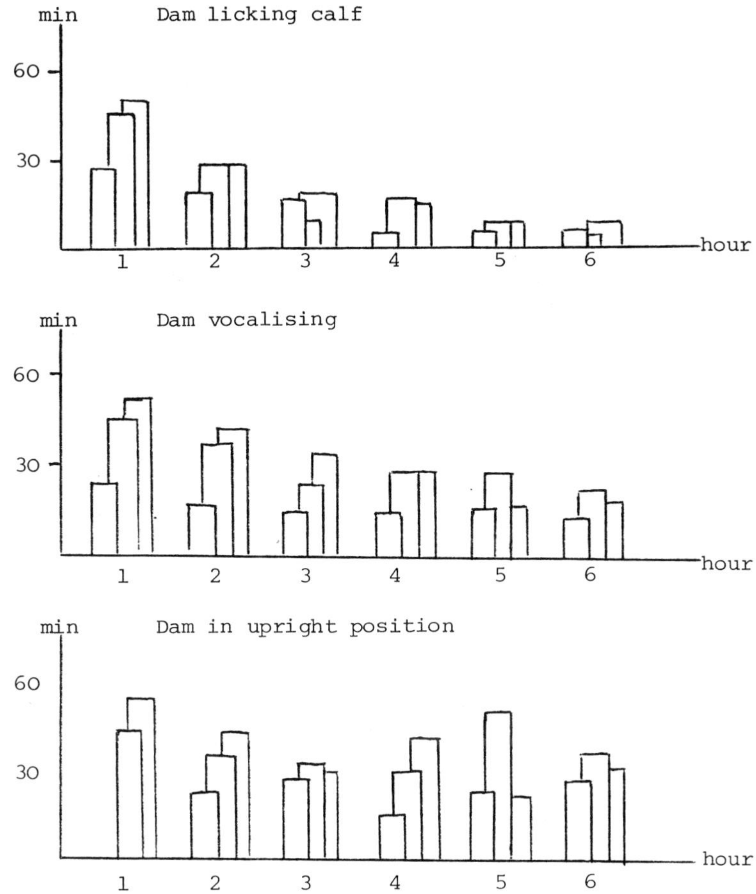

Fig. 1. First column in every triplet: D-heifers (n = 4, 1 - 6 h)

Second column in every triplet: D-cows (n = 6, 1 - 3 h and n = 4, 4 - 6 h)

Third column in every triplet: B-D-cows (n = 8, 1 - 3 h and n = 7, 4 - 6 h)

The cow was oriented towards the calf as the latter initiated teat-seeking, and remained for the most part in a stationary position during the calf's teat-seeking and suckling, turning the head to continue licking the calf. During suckling the calf stood in an opposite parallel position to the cow, the cow licking the hindquarter of the calf, including the anogenital region. During the cow's licking of the anogenital region, three calves passed meconium and in two cases this was carefully licked up directly from the anus.

Placenta was expelled 230 (± 79) min after parturition, and the cows were eating placenta for a longer continuous period of 26 (± 11) min.

As soon as the calf was born and the mother got up and began licking it, she became protective of the area where the calf was, i.e. she kept a territory around herself and the calf. Other cows and calves coming close could either be threatened or even attacked and butted away.

All of the BD mother-young pairs, except one, remained on the area where the calf had been born until the calf had got up, and had suckled the cow. This period lasted for at least the first two hours after parturition.

Isolation was artificially made in the case of the two housed groups, D-cows and D-heifers, by placing them in a calving box.

The four D-cows calving without assistance completed parturition while lying on the side. These cows were slower to rise after parturition than the BD-cows, as they rose from 1 to 6 min after parturition but then immediately started licking the calves.

Apart from the first activities of the D-cows being late compared to the BD-cows, we could not from our results show significant differences in the behaviour characterising maternal behaviour, such as licking of the calf and vocalising. Like the BD-cows, the D-cows accepted the calf's teat-seeking and suckling, and placentophagy was pronounced within both groups of cows.

All the D-heifers needed assistance at parturition and remained recumbent for 63 - 318 min. The newborn calves were placed immediately in front of the dams, which then - in 3 of the 4 cases - started licking the calf.

The D-heifers differed from both groups of cows, especially from the BD-cows. Thus we found significant differences and pronounced tendencies showing that the D-heifers stood up less and vocalised less during the first six hours and licked the calf less during the first hour after parturition. Yet we found only a few significant differences between heifers and cows with respect to licking the calf and vocalising, but the licking and vocalisation were not so intensively and continuously carried out by the heifers as they were by the cows observed.

In two cases, the heifers showed an abnormal maternal behaviour in that one heifer rejected the calf by butting it away with the head and another ignored the calf. None of the heifers immediately accepted the calf's teat-seeking advances; the dam could move so as to remain facing the calf, thereby preventing it from reaching the udder. All the heifers, however, accepted the calf and allowed suckling within the observation period.

All the D-heifers showed placentophagy, but the period between the expulsion of placenta and the beginning to eat it was considerable (17 - 120 min), as compared to the cows, all but one of which started eating within one minute of expulsion.

BEHAVIOUR OF THE CALF FROM BIRTH TO THE INITIAL PHASE OF SUCKLING

When the calf was born, it was lying stretched out on one side. From this position the calf rose to sternal recumbency, and the first activities were then mainly attempts to rise to its feet.

As mentioned above, we could not from our results conclude that there was a difference in the intensity of licking between the two groups of cows. But if there was a difference it apparently did not influence the time at which the calf was upright for the first time. Thus we could not show any difference between the BD-cows' and the D-cows' calves with respect to the time they rose to sternal recumbency, the time they attempted to stand for the first time, and the interval from this to their being upright for the first time (Table 1). But the D-cows'

TABLE 1

THE FIRST ACTIVITIES OF THE CALF

	Interval (min) from birth to the calf rising into sternal recumbency		Interval (min) from birth to the calf attempting to stand for the first time		Number of attempts to stand (No. of min within which attempts to stand have been recorded)		Interval (min) from the calf attempting to stand for the first time to the calf being upright for the first time		Interval (min) from the calf being upright for the first time to the calf standing for the first time		Interval (min) from birth to the calf standing for the first time		Interval (min) from birth to the calf vocalising for the first time	
	$\bar{x}$	SD	$\bar{x}$	SD	$\bar{x}$	SD	$\bar{x}$	SD	$\bar{x}$	SD	$\bar{x}$	SD	$\bar{x}$	SD
BD-cows' calves n = 8	$<2 \leqslant x_i \leqslant 18$		14.0	9.7	6.4	2.7	22.0	19.5	4.1	3.6	40.1	18.5	216.8*	108.1*
D-cows' calves n = 6	8.3	5.1	11.7	7.1	11.2	5.0	42.5	34.4	13.3	7.6	67.5	34.6	51.8	67.7
D-heifers' calves n = 4	22.0	18.3	32.8	23.0	7.8	1.0	56.3	30.4	15.8	22.9	104.8	46.2	15.3	18.3

* n = 6 in the computed mean (nos. 148 and 108 not included); n = 8 in the M-W tests for BD-cows' calves

TABLE 1 (cont.)

	Interval (min) from birth to the calf rising into sternal recumbency	Interval (min) from birth to the calf attempting to stand for the first time	Number of attempts to stand (No. of min within which attempts to stand have been recorded)	Interval (min) from the calf attempting to stand for the first time to the calf being upright for the first time	Interval (min) from the calf being upright for the first time to the calf standing for the first time	Interval (min) from birth to the calf standing for the first time	Interval (min) from birth to the calf vocalising for the first time
	$\bar{x}$ SD	$\bar{x}$ SD	$\bar{x}$ SD	$\bar{x}$ SD	$\bar{x}$ SD	$\bar{x}$ SD	$\bar{x}$ SD
M-W tests:							
BD-cows' calves - D-cows' calves	-	-	p = 0.030	-	p = 0.015	p = 0.091	p < 0.006*
BD-cows' calves - D-cows' calves	p < 0.036	p = 0.024	-	p = 0.014	-	p = 0.004	p = 0.014*
D-cows' calves - D-heifers' calves	p = 0.057	p = 0.057	p = 0.057	-	-	p = 0.086	-

* n = 6 in the computed mean (nos. 148 and 108 not included); n = 8 in the M-W tests for BD-cows' calves

calves made more attempts than the BD-cows' calves before they were upright ($p = 0.030$), and the interval from being upright to standing was also larger for this group ($p = 0.015$). This difference could be traced back to a factor other than maternal behaviour. The floor in the calving-box tended to be very slippery from birth fluids, so that the D-cows' calves had difficulties in rising and remaining on their feet.

The D-heifers' calves were later in performing their first activities, especially when compared to the BD-cows' calves. Differences in time were significant, both for rising to sternal recumbency ($p = 0.036$), attempting to rise for the first time ($p = 0.024$), and at being upright for the first time ($p = 0.014$). But the total number of attempts to stand before being upright, and the interval from this to standing for the first time was no larger for the D-heifers' calves than for the BD-cows' calves. This may have been caused by the fact that as none of the heifers rose earlier than their calves, the floor of the calving box was consequently not so slippery and bare as was often the case when the D-cows' calves attempted to stand.

It is not possible to conclude whether the delayed activities of the heifers' calves were caused by the effects of difficulties at parturition or insufficient maternal behaviour.

When looking at the interval from birth to standing for the first time, we found that the BD-cows' calves were the fastest to stand (Table 1), the D-cows' calves being next, and the D-heifers' calves being slowest. The difference was only significant ($p = 0.004$) between the first and the last group, and we found only tendencies to difference ($p = 0.091$ and $p = 0.086$) elsewhere.

With respect to the interval from birth to vocalising for the first time, we found the opposite order in the three groups of calves, with the BD-cows' calves vocalising significantly later than both the D-cows' and the D-heifers' calves ($p = 0.006$ and $p = 0.014$). Whether these differences were caused by a difference in maternal behaviour, breed, or environment, could not be determined.

BEHAVIOUR OF THE CALF DURING TEAT-SEEKING AND SUCKLING

Shortly after the calves were standing they initiated teat-seeking at their dams (Table 2). They oriented their teat-seeking towards places that were about the level of their heads when standing, mostly at the lower outline of dam. During teat-seeking the calf nosed, nudged, or made forceful pushing head movements at the dam.

Even if the calves generally directed their teat-seeking at places between thorax and abdomen, it was seen, especially among the BD-cows' calves, that they spent more time teat-seeking on the udder and the teats than on other parts of the body. In some cases, teat-seeking was exclusively directed at the udder region. Contact with the udder itself therefore seems to have a stimulus value for the neonate.

The D-cows had big and lowhanging udders compared to the other two groups. This influenced the pattern and time spent actively teat-seeking (i.e. the number of minutes in which teat-seeking was observed before locating a teat). As mentioned above, the calves oriented their teat-seeking activity towards places about the level of their head when standing. The teats of the BD-cows and the D-heifers were about this height, while the teats of the D-cows were considerably lower. As a consequence, the teat-seeking of the D-cows' calves was carried out high above the teats.

With respect to the distribution of the teat-seeking at the fore-, middle-, and hindquarter of the dam, we found this to be the same for all but one of the BD-cows' calves, namely 19% at the fore- and middlequarter respectively and 62% at the hindquarter. Teat-seeking was thus mostly concentrated at the hindquarter and here especially at the udder region.

When we looked at the D-cows' calves, there was not the same regularity with respect to the distribution of the teat-seeking, but rather most search was directed at the middle- and hindquarter of the dam. As was the case with the BD-cows' calves, teat-seeking of the D-cows' calves directed at the hindquarter was mostly in the udder region, but because the teats were situated very low, the calf sought for a long time without locating a teat. This eventually resulted in a lot of

TABLE 2

TEAT-SEEKING

	Interval (min) from birth to when teat-seeking directed at the dam is initiated.			Interval (min) from when the calf stands for the first time to when teat-seeking directed at the dam is initiated.			Time spent actively teat-seeking prior to the 1st suckling spell (No. of min, within which teat-seeking directed at the standing dam has been recorded).			Interval (min) from when teat-seeking directed at the dam is initiated to when the calf is suckling for the first time.		
	n	$\bar{x}$	SD	n	$\bar{x}$	SD	n	$\bar{x}$	SD	n	$\bar{x}$	SD
BD-cows' calves	8	43.1	19.6	8	3.0	3.7	8	10.5	13.2	7*	17.0	25.6
D-cows' calves	6	78.0	32.7	6	10.5	11.4	4	33.3	29.7	4	127.0	92.9
D-heifers' calves	4	131.8	43.4	4	12.0	10.0	4	9.0	3.4	4	89.8	72.6
<u>M-W tests</u>:												
BD-cows' calves - D-cows' calves	$p = 0.041$			-			$p = 0.033$**			$p = 0.006$		
BD-cows' calves - D-heifers' calves	$p = 0.002$			$p = 0.055$			-			$p < 0.036$		
D-cows' calves - D-heifers' calves	$p = 0.033$			-			$p = 0.056$**			-		

* No. 108 omitted, as the calf had no possibility of suckling prior to the 4th hour.

**n = 5 in the M-W tests for D-cows' calves.

search being directed elsewhere, at the middlequarter. One of the D-cows' calves directed most teat-seeking around the fore-legs of the cow. In this case, the udder and teats were extremely lowhanging.

The D-heifers' calves did not show any regular pattern in their teat-seeking. This must be seen in connection with the fact that the D-heifers did not stand for the calves' teat-seeking advances as opposed to the other two groups of dams who did.

The D-cows' calves spent significantly more time actively teat-seeking than did the BD-cows' calves ($p = 0.033$), and there was a pronounced tendency in the same direction when compared to the D-heifers' calves ($p = 0.056$). Between the BD-cows' calves and the D-heifers' calves we found no difference (Table 2).

All BD-cows' calves but two sought for a teat with interruptions of 2 min at most before locating a teat and suckling. These calves spent from 1 to 17 min from the initiation of teat-seeking to the initiation of suckling.

Teat-seeking of the D-cows' calves was not concentrated in one continuous period. Thus all the calves had at least one period lying down before succeeding in suckling. These interruptions could be caused by repeated, unsuccesful attempts of locating a teat and suckle. The 4 D-cows' calves that suckled within the first 6 h spent from 22 to 248 min from the initiation of teat-seeking to the initiation of suckling.

As was the case with the D-cows' calves, the D-heifers' calves spent significantly longer time from the initiation of teat-seeking to suckling than BD-cows' calves ($p = 0.036$). This was caused by the behaviour of the dams which in 3 of the 4 cases were lying down for long periods.

All but one of the BD-cows' calves suckled within 2 h of parturition. The calves in the other two groups clearly suckled later. Thus the D-cows' calves that suckled within the first 6 h (4 out of 6) did so in 3 - 5 h, and the D-heifers' calves in 3 - 6 h. We found no difference between these two groups.

TABLE 3

SUCKLING

	Interval (min) from birth to when the calf is suckling for the first time.			Duration of the suckling spell (min)						Suckling time (No. of min within which suckling has been recorded during the suckling spell).					
				1. Suckling spell			2. Suckling spell			1. Suckling spell			2. Suckling spell		
	n	x̄	SD	n	x̄	SD	n	x̄	SD	n	x̄	SD	n	x̄	SD
BD-cows' calves	7*	60.9	18.0	8	31.1	12.4	4	9.8	3.4	8	22.9	7.6	4	9.5	3.4
D-cows' calves	4	203.3	60.6	3**	37.0	14.7	1	31		3**	22.0	12.2	1	9	
D-heifers' calves	4	221.5	99.5	3***	29.0	13.7	2	21.0	14.1	3***	17.7	4.7	2	16.0	11.3
M-W tests															
BD-cows' calves - D-cows' calves		p = 0.003			-			.			-			.	
BD-cows' calves - D-heifers' calves		p = 0.003			-			.			-			.	
D-cows' calves - D-heifers' calves		-			-			.			-			.	

* No. 108 omitted, as the calf had no opportunity for suckling prior to the 4th hour.

** No. 152 omitted, as the 1. suckling spell was not terminated within the observation period.

*** No. 184 omitted, as the 1. suckling spell was not terminated within the observation period.

The mean suckling time of all calves in the first suckling spell* was 21.6 (±7.9) min. No differences could be shown between the three groups either with respect to the length of the first suckling spell or the suckling time in the first suckling spell. Seven calves had a second suckling spell with a mean suckling time of 11.3 (±6.1) min, which was shorter than the first suckling spell.

During suckling all calves stood in an opposite parallel position to the dam. Butting head movements directed at the dam were also observed during teat-seeking. Tail-shaking was not pronounced within the observed calves.

DISCUSSION AND CONCLUSIONS

Selman et al. (1970) concluded from their observations that there were differences in maternal behaviour between beef cows and dairy cows/heifers. Although the BD-cows were not pure-bred beef cows one might expect the differences between the BD-cows and the D-cows with respect to breed, rearing conditions, and environments to influence the maternal behaviour of the cow. Apart from the fact that the BD-cows responded quicker to their newborn calves than the D-cows, we could not conclude from our results that there were differences in maternal behaviour between the two groups of cows.

It must be mentioned, however, that a recording per minute instead of a recording within every minute of the cow's licking and vocalisation would be a better method for showing differences in the frequency of licking and vocalisation.

The maternal behaviour of the D-heifers differed from both groups of cows. This was most probably caused by a combination of difficulties at parturition, and lack of maternal experience (Alexander, 1960; Arnold and Morgan, 1975; Sharafeldin and Kandeel, 1971 - all about sheep; Donaldson, 1970).

The different types of behaviour the cow exhibits after parturition, such as licking the newborn and vocalising, function in different ways as stimuli for the calf, and influence its condition.

*The division into suckling spells was made by an assessment of the data. According to this the longest interruption of suckling in a suckling spell was 10 min, and the shortest interval between two suckling spells was 21 min.

The importance of the dams licking of the offspring is mentioned by many authors (Barron, 1954; Ewer, 1968; Sambraus, 1973). Barron (1954) writes about kids:

> 'I have the impression that the drying, the licking, and the grooming are important in raising the general level of neural excitability of the kid, and thereby hasten his ability to rise on his knees, to orient himself, and to stand.'

Selman et al. (1970) found that beef cows' calves stood significantly earlier than dairy cows' calves, and it was established that the beef cows carried out their grooming activities more vigorously, and for a longer period than most of the other dams.

It was not possible from our observations to find any correlation between licking activity by the dams and standing times of the calves either within or between the two groups of cows and calves.

Between the BD-cows' calves and the D-cows' calves respectively and the D-heifers' calves we found significant differences and pronounced tendencies respectively with respect to standing times. These differences could be related both to difficulties at the birth of the calves of the D-heifers, and differences in maternal stimuli from cows and heifers including, among other things, insufficient licking by the heifer.

The localisation of a teat by the newborn calf has been described as trial and error learning (Fraser, 1968; Hafez and Lineweaver, 1968; Stephens and Linzell, 1974 - about kids). Hafez and Lineweaver (1968), however, write that the young calf spends more time nuzzling an area or object which feels like the cow's teat than it will spend on body parts which are covered by a heavy hair coat.

The localisation of a teat therefore seems to be achieved by trial and error learning, where stimuli from tactile contact with the udder itself influence the behaviour of the calf so that orientation at the dam is more and more directed towards the udder region. This would be in agreement with our findings, especially within the group of BD-cows' calves.

Selman et al. (1970) found that the shape of the dam's underbelly had an effect upon where on the dam's body the calf concentrated its teat-seeking activities, and that the shape had a significant effect upon the total teat-seeking time prior to first suckling.

We found that Selman et al's division of the cows/heifers into 'good' (udder and teats on the level with or higher than the dam's xiphisternum) and 'poor' shape (xiphisternum was the highest part) was too sharp because it implies that the calf primarily directs its teat-seeking at the highest part of the dam's underbelly, and the possible stimulus value of the udder itself had not been taken into account. Moreover, the effect of differences in maternal behaviour upon the calf's teat-seeking behaviour had also been disregarded. We therefore chose to maintain our division into three groups, and look at differences between these.

The big, lowhanging udder of the D-cows caused the calves to spend significantly longer time actively teat-seeking than the BD-cows' calves and the D-heifers' calves, and the long time spent searching without locating a teat made the calves pause. That teat-seeking is subject to fatigue in lambs is described by Alexander and Williams (1966).

While Selman et al. (1970) and Donaldson (1970) found that calves of dams with low-positioned udder and teats directed a great deal of their teat-seeking around the front legs of their dams, this was only pronounced in one of the D-cows' calves.

Although the D-heifers did not immediately accept their calves' teat-seeking advances, this was not an impeding factor, when compared to the udder-position of the D-cows, with respect to the time spent actively teat-seeking before locating a teat. But as was the case with the D-cows' calves, the D-heifers' calves spent a long time from the initiation of teat-seeking to suckling for the first time. This was caused by the fact that the dams were lying down for long periods.

The BD-cows' calves spent a short time actively teat-seeking before locating a teat, and the interval from the initiation of teat-seeking to suckling was shorter than within the other two groups. This was most likely because there were no impeding factors within this group, but rather a combination of a good udder-position and maternal acceptance of the calf's teat-seeking.

The BD-cows' calves suckled significantly faster than the other two groups of calves (only those D-cows' calves which did suckle during the observation period (4 out of 6) were considered). This was caused both by the fact that the BD-cows' calves initiated their teat-seeking activity earlier than the other two groups of calves, and that the interval from this to suckling was significantly shorter for this group as well.

While one could be fairly certain that the BD-cows' calves would obtain colostrum in very early life when left alone with their dams this did not seem to be sufficient within the group of D-cows' calves and D-heifers' calves.

During suckling the calf stood in an opposite parallel position to the dam. On account of this position it is possible for the dam to continue licking its calf. Licking of the ano-genital region may function as a stimulus for defaecation and urination (Ewer, 1968; Hafez and Lineweaver, 1968; Walker, 1950), and during the licking and sniffing the dam has an opportunity of identifying its offspring (Poindron and Neindre, 1975; Sambraus, 1971; Schloeth, 1958). Finally this position ensures tactile contact between mother and calf.

REFERENCES

Alexander, G., 1960. Maternal behaviour in the Merino ewe. Proc. Aust. Soc. Anim. Prod. 3, 105-114.

Alexander, G. and Williams, D., 1964. Maternal facilitation of sucking drive in newborn lambs. Science 146, 665-666.

Alexander, G. and Williams, D., 1966. Teat-seeking activity in lambs during the first hours of life. Animal Behaviour 14, 166-176.

Arnold, G.W. and Morgan, P.D., 1975. Behaviour of the ewe and lamb at lambing and its relationship to lamb mortality. Applied Animal Ethology 2 (1), 25-46.

Arthur, G.H., 1961. Some observations on the behaviour of parturient farm animals with particular reference to cattle. Veterinary Review 12 (3), 75-84.

Bareham, J.R., 1976. The behaviour of lambs on the first day after birth. British Veterinary Journal 132 (2), 152-162.

Barron, D.H., 1954. Proceedings of the 1st Conference on Group Processes, New York, 1954, p. 221-258.

Blauvelt, H., 1954. Proceedings of the 1st Conference on Group Processes, New York, 1954, p. 221-258.

Brownlee, A., 1950. Studies on the behaviour of domestic cattle in Britain. Bulletin of Animal Behaviour 8, 11-20.

Collias, N.E., 1956. The analysis of socialisation in sheep and goats. Ecology 37, 228-239.

Donaldson, S.L., 1970. The effects of early feeding and rearing experiences on social, maternal, and milking parlor behaviour in dairy cattle. Thesis, Purdue University, West Lafayette, Indiana.

Edwards, S.A. and Broom, D.M., 1979. The period between birth and first suckling in dairy calves. Research in Veterinary Science 26 (2), 255-256.

Ewer, R.F., 1968. Ethology of mammals, kap. 5. London 1968.

Finger, K.H. and Brummer, H., 1969. Beobachtungen über das Saugverhalten mutterlos aufgezogener Kälber. Deutsche Tierärztliche Wochenschrift 76, 665-667.

Fraser, A.F., 1968. Reproductive behaviour in ungulates, kap. 7-8. London and New York, 1968.

George, J.M. and Barger, I.A., 1974. Observations of bovine parturition. Proc. Aust. Soc. Anim. Prod. 10, 314-317.

Hafez, E.S.E., 1964. Some physiological and behavioral responses of the neonate. Cornell Veterinarian 59, 545-560.

Hafez, E.S.E. and Bouissou, Marie-France, 1975. The behaviour of cattle. In: E.S.E. Hafez (Ed.): The behaviour of domestic animals. London 1975, p. 203-245.

Hafez, E.S.E. and Lineweaver, J.A., 1968. Suckling Behaviour in Natural and Artificially Fed Neonate Calves. Zeitschrift für Tierpsychologie 25, 187-198.

Hersher, L., Richmond, J.B. and Moore, A.U., 1963. Maternal behaviour in sheep and goats. In: H.L. Rheingold (Ed.): Maternal behaviour in mammals, New York 1963, p. 203-231.

Kiley, M., 1973. The vocalizations of ungulates, their causation and function. Zeitschrift für Tierpsychologie 31 (2), 171-222.

McBride, G., Arnold, G.W., Alexander, G. and Lynch, J.J., 1967. Ecological aspects of the behaviour of domestic animals. Proceedings of the ecological Society of Australia 2, 133-165.

Poindron, P. and Le Neindre, P., 1975. Comparaison des relations mère-jeune observées lors de la tétée chez la brebis (Ovis aries) et chez la vache (Bos taurus). Annales de Biologie Animale, Biochemie, Biophysique 15 (3), 495-501.

Sambraus, H.H., 1971. Zum Mutter-Kind-Verhalten der Wiederkäuer. Berliner und Münchener Tierärztliche Wochenschrift 84, 42-47.

Sambraus, H.H., 1973. Zum Beitrag "Sauberlecken oder nicht" von H. Bielenberg. Praktische Tierarzt 54 (3), 108.

Schloeth, R., 1958. Uber die Mutter-Kind-Beziehungen beim halbwilden Camarque-Rind. Saugetierkundliche Mittelungen 6, 145-150.

Selman, I.E., McEwan, A.D. and Fisher, E.W., 1970. Studies on natural suckling in cattle during the first eight hours *Post partum*. I. Behavioural studies (Dams), II. Behavioural studies (Calves), Animal Behaviour 18, 276-289.

Selman, I.E., McEwan, A.D. and Fisher, E.W., 1971. Studies on Dairy Calves Allowed to Suckle their Dams at Fixed Times *Post Partum*. Research in Veterinary Science 12 (1), 1-6.

Sharafeldin, M.A. and Kandeel, A.A., 1971. Post-lambing maternal behaviour. Journal of Agricultural Science 77, 33-36.

Smith, F.V., 1965. Instinct and learning in the attachment of lamb and ewe. Animal Behaviour 13, 84-86.

Stenbæk, B., Bech Andersen, B. and Bülow-Olsen, A., 1980. Ammekøer og landskabspleje (racer, produktion, management, botanik). Beretning fra Statens Husdyrbrugsforsøg nr. 494.

Stephens, D.B. and Linzell, J.L., 1974. The development of sucking behaviour in the new-born goat. Animal Behaviour 22 (3), 628-633.
Walker, D.M., 1950. Observations on behaviour in young calves. Bulletin of Animal Behaviour 8, 5-10.

DISCUSSION

P. Le Neindre *(France)* Do the multiparous dairy cows have normal relations with the calves during preceding lactations, and at the beginning, at calving?

Else Hermann *(Denmark)* Yes, all the dairy cows except the two dairy cows that needed assistance at parturition.

J.P. Signoret *(France)* The question is whether or not your dairy cows had contact with the calves. In your experiment the calves were kept with the cows for 72 h. Was this similar to earlier calvings or were the dairy cows separated from the calves any earlier or later?

Else Hermann The situation was the same as in previous calvings.

J.P. Signoret Could you tell us the breed of your BD cows, your dairy cows and the breed of the calves in each case?

Else Hermann The dairy cows were Black and White Holstein Friesians and the beef dairy cows were Aberdeen Angus and crossed Blond d'Aquitaine and Charolais.

J.P. Signoret They belonged then to different crosses, and were the calves of the same breed as the sires?

Else Hermann No, they were all mixed up.

R. Moss *(UK)* Was there within any of the three groups any significant variation in the time factors that you were recording? For instance, you recorded 89 minutes before suckling after the calves had found the teats. Was there within that group one calf that suckled very quickly? Was there any vast difference in the figures or do the means that you show represent something significant?

Else Hermann Within the group of BD cows-calves the figures were very close but within the other groups there were some considerable differences.

J.Y. Blom *(Denmark)* This may not be quite fair to you but have you considered the fact that all the heifers had their calves pulled out at birth? How do you think this influenced the times?

Else Hermann This must have influenced the times quite considerably. However, I have observed heifers that have not had calves pulled out and their times differed greatly as well.

J.Y. Blom So you have looked at 'normal' heifers and have found the same results.

J.P. Signoret My question concerns variability amongst heifers and I would like to make reference to a study carried out in France on maternal behaviour in primiparous ewes. Some primiparous ewes are either completely unadapted, rejecting and even attacking their newborn lambs, whilst others are immediately normal. Therefore, the variability in primiparous ewes is very wide since there are either completely normal or completely abnormal ewes. Therefore the variability is not a good index.

Have you observed heifers immediately normal as multiparous cows and also totally abnormal heifers, or is yours a true variability? I know that it is difficult to have large figures when making such long observations, but have you observed cases of total inadaptation and total adaptation?

Else Hermann I have not observed such behaviour in this study but I have seen heifers immediately accept calves after parturition. All the heifers in this case were abnormal in their maternal behaviour.

Nanny Stenum *(Denmark)* There were differences in the degree of acceptance.

P.J. O'Connor *(Ireland)* In relation to the action of the cow licking the calf, did you observe at any time calves who suckled without being licked?

Else Hermann Yes, I have observed this.

P.J. O'Connor So, in other words, the cow accepted the calf and allowed suckling without the contact of licking the calf?

Else Hermann No, I was thinking rather of a calf suckling another dam than its mother.

COW-CALF RELATIONSHIPS : THE EFFECT OF MANAGEMENT SYSTEMS

P. le Neindre

Laboratoire de la Production de Viande,
Centre de Recherches de Clermont-Ferrand, INRA,
CRZV de Theix, 63110 Beaumont, France.

INTRODUCTION

From a behavioural point of view it can be seen that there are three main systems of calf management. These are:

1. The calf is kept with the mother, mostly at pasture during the grazing season. This is the normal management system used for beef breeds such as the Charolais and the Limousin and for hardy breeds such as the Salers.
2. The calf is separated from its mother and is weaned early in order to produce future reproductive animals or beef animals. The calves are generally bucket-fed and are soon divided into groups of animals of the same age.
3. There is also the system in which calves are milk fed until slaughter at three or four months for veal production. This system is generally used with calves from dairy or dual-purpose breeds.

Rearing of French calves is divided reasonably evenly between these three systems.

In order to analyse the effect of these different management systems on the behaviour and survival of calves we will first present results from observations of the cow-calf relationship in herds where there was a minimum of disturbance, and we will then analyse the adaptability of animals under the types of management systems considered.

The majority of results presented in this paper were drawn from two breeds: the Salers, a hardy breed where cows are used as suckling cows, and the dairy Friesian.

COW-CALF RELATIONSHIPS WITHIN A HERD

First of all the relationship between a mother and her calf is described, and then the relationship between the mother-offspring pair and other animals in the herd.

Thirteen cows who calved at pasture were observed in order to analyse the mother-offspring relationship immediately after birth. Seven of the cows were Salers with Salers calves and six were Friesian cows whose offspring were crossbred Salers Friesians. Each mother-offspring dyad was observed for 4 h after birth. All cows calved without assistance. The results are summarised in Table 1.

TABLE 1

SOME CHARACTERISTICS OF MOTHER-OFFSPRING BEHAVIOUR IN THE 4 h AFTER BIRTH (13 CALVINGS AT PASTURE, 7 SALERS AND 6 FRIESIANS)

	Mean	Range
Number of dyads	13	
<u>Calf</u>		
Interval (min) between birth and:		
First head-lifting	3.3	1 - 10
First attempt at standing	9.4	4 - 19
First standing up	51.1	32 - 83
First sucking	72.6	41 - 132
Total sucking time (min)	25.7	12 - 44
<u>Mother</u>		
Interval (min) between birth and:		
Standing up	3.3	1 - 12
Onset of licking	6.0	1 - 15
Total licking time (min)	43.0	18 - 72

The cows stood up soon after calving and licked their young. The mean point at which the calves tried to stand up was 9.4 min after birth and they succeeded 49 min after that first attempt. They then required only a short time in which to find the udder and begin sucking (Table 1). There was considerable variation in these results, but all the calves suckled within the 4 h period and whilst the mean results are very much shorter times than those of Hafez (1969) and Edwards and Broom (1979), they are similar to those obtained by Selman et al. (1970).

The mother-offspring relationships rapidly became structured. The criteria of suckling, licking and cow-calf proximity are used to describe the mother-offspring relationship.

Suckling was usually performed in the parallel-inverse position, particularly at the beginning of the suckling period. The calf can also suck when in a perpendicular or rear position, however, particularly at the end of the suckling period (Poindron and Le Neindre, 1975). As the calf grows older, the number of sucklings decreases (Wagnon, 1960; Ewbank, 1969; Petit, 1972; Le Neindre, 1975; Nicol and Sharafeldin, 1975) (Figure 1).

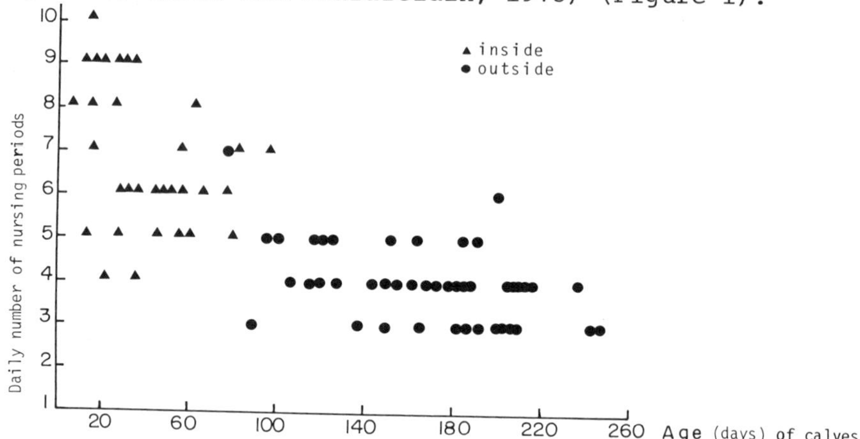

Fig. 1. Number of nursing periods/day according to the age of the calves

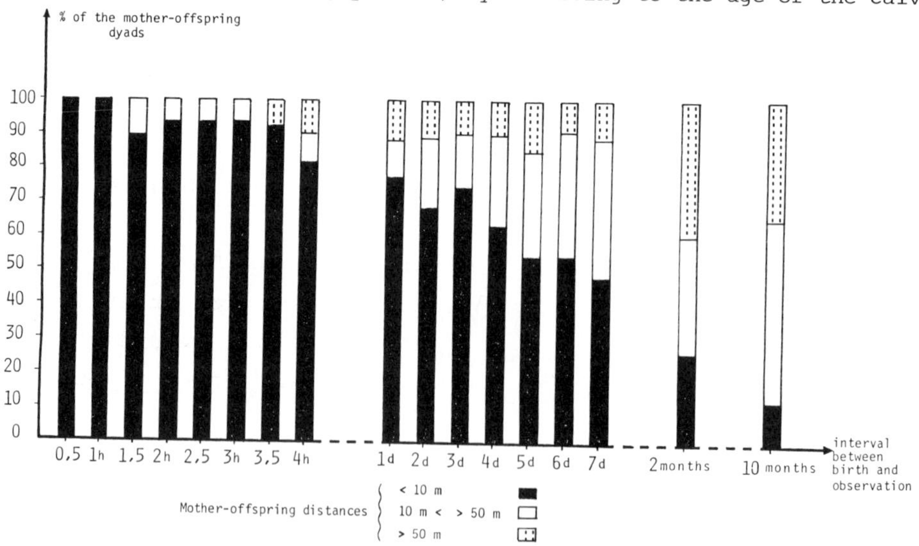

Fig. 2. Distance between mother and calf in relation to the calf's age (20 dyads)

The calf passed in front of the cow before or during 29% of suckling periods (Le Neindre and Garel, 1977). This is a lower frequency than that which has been observed in the ewe, which indicates that olfaction is not as important for the cow as it is for the ewe (Poindron and Le Neindre, 1975).

There was also very frequent licking and this activity occupied 43 min during the first 4 h after birth. This incidence persisted until weaning.

Mother-offspring proximity varied during the day. The proximity was measured at intervals of 30 min during the first 4 h after birth, then twice a day for seven days, at two months and at ten months. In the period immediately following birth the mothers remained with their young. Some days later the distance between the mother and her offspring began to increase and continued to do so until the calf was ten months of age (Figure 2).

The cow and her offspring may also be involved with other cows and calves in the herd, but interactions outside the mother-offspring dyad are rare (Figure 3) (Le Neindre and Garel, 1977). Some cows licked alien calves but often ended this activity by butting the calf, whilst some calves tried, unsuccessfully, to suck alien cows.

The distance between the cow and the herd before and after calving was measured (Figure 4). Before calving, the mother remained in the herd. Just after calving, the other cows often came quite close to the dyad. The postparturient cow remained with her calf outside the herd for a few days and then rejoined the herd.

The cow is capable of protecting her young from predators. Although this is a difficult attribute to quantify, it is important for the herdsman, who is apparently often considered a predator by the postparturient cow, especially cows of beef breeds.

The interactions described do not disappear totally at weaning. The integration of the heifer into the adult group may be influenced by the persistence of the bond with the dam, and later various patterns of adult behaviour can be influenced by experience during development. This has been shown for sexual

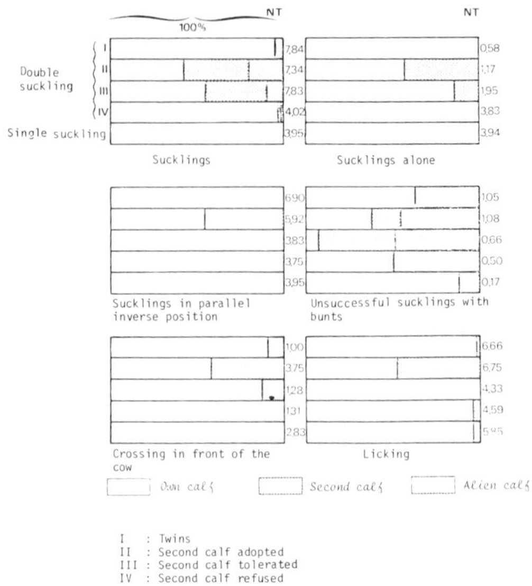

Fig. 3. Frequency of acts observed between cows and calves. (NT : Total number of interactions per day and per cow)

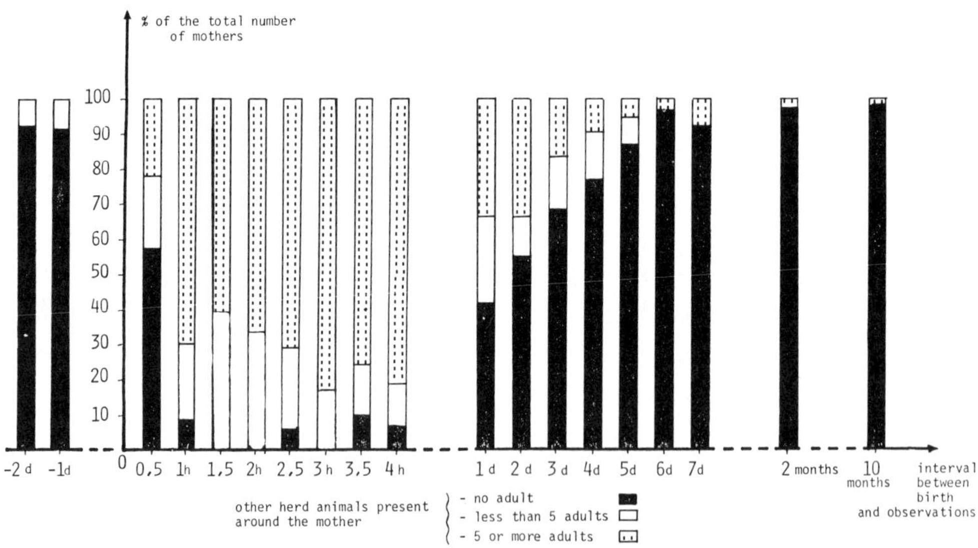

Fig. 4. Mother - herd distance before and after calving (20 cows)

and maternal behaviour in laboratory animals but has yet to be studied in domestic cattle. The only aspect which has been investigated as yet is the importance of the social rank of the cows in determining the social rank of their calves. This was done by calculating the relationship between the social dominance of 23 dyads of cows and that of 23 dyads of calves nursed by them. No effect was found.

MODIFICATION OF THE NORMAL COW-CALF RELATIONSHIP

Given the normal pattern of interaction between cow and calf, it is possible to investigate the consequences of changes which result from modern management techniques. This study pays particular attention to the double suckling technique, where a cow has to suckle an alien calf, and to artificial feeding, where the calf is isolated from the mother.

Double suckling system

When a cow has a high potential for milk production she can often rear two calves. This poses no problem when the cow gives birth to twins, but it is much more difficult to obtain a normal relationship between a cow and an alien calf (Figure 3).

TABLE 2

COW-CALF RELATIONSHIPS IN THE DOUBLE SUCKLING SYSTEM AND INFLUENCE ON THE GROWTH OF THE TWO CALVES DURING THE GRAZING SEASON (141 DAYS)

Cow-calf relationships	Number of cows	Daily liveweight gain (g/day)	
		Mother's calf	Second calf
Twins	26	991 x 2	
Second calf adopted	26	919	925
Second calf tolerated	43	1 040	784
Second calf refused	16	1 118	714

When the second calf was not adopted, it did not suck in the parallel-inverse position and it was not licked. It was however able to suck in a position other than the parallel-inverse while the mother's own calf was sucking. In this case it can be said that the calf was 'tolerated'. When the calf had no relationship with the cow, it was said to have been refused (Le Neindre and Garel, 1977). The liveweight gain of the two calves was dependent on their relationship with the mother. When the second calf was adopted, the liveweight gain of both calves was the same, whereas the poorer the second calf's relationship with the mother, the lower its growth rate and the more the mother's own calf grew (Table 2) (Le Neindre et al., 1978).

Calves reared without the mother

About one third of calves in France are reared with their mother and so can develop normal behavioural interactions such as those which have been described. The remainder are isolated from the mother and reared either alone or with calves of the same age. The influence of this management system on calf behaviour, particularly in the period immediately after birth, has been studied.

The observations were focussed on the Salers and Friesian breeds. The aim of this comparison was to test the hypothesis that a production system such as dairying creates the equivalent of an ecological niche which involves special nutritional, physiological and behavioural conditions. The domestic breed designated for such a production system may have undergone a kind of evolution not only in milk production itself, but also in the particular traits pertinent to this ecological niche.

The activity of calves during the first 4 h after birth was observed. Sixteen Salers were studied, 8 with their mothers and 8 without, as were 8 isolated Friesians. The isolated Salers were less active than those who were with their mothers, whilst the activity shown by the isolated Friesians did not differ greatly from that shown by the mothered Salers (Table 3) (Le Neindre et al., 1979).

Colostrum intake, which is very important for calf survival, was also measured. Salers who sucked their own mothers showed a greater colostrum intake than those Salers which were bottle fed (Table 4), whilst Friesian calves drank more bottle-fed colostrum than the Salers (Figure 5).

The aptitudes of the two breeds for drinking from a bucket were then compared, with clear cut results. All the Friesians drank, but few Salers calves were able to drink from the bucket

(Figure 6). Similarly, Friesian calves are known to be able to eat more solid feed than beef calves.

TABLE 3

PRINCIPAL BEHAVIOURAL TRAITS OF FRIESIAN AND SALERS CALVES DURING THE 4 h PERIOD AFTER BIRTH

	Salers		Friesian
	With mother	Alone	Alone
Number of calves	8	8	8
Interval (min) between birth and first assumption of intermediate position	17.3[a]	71.2[b]	29.5[a]
Standing position:			
Number of calves to stand	7	2	6
Interval (min) between birth and standing up (=240 if the calf did not stand up)	76.0[a]	228.8[b]	153.5[a]
$\frac{\text{Time spent in each position}}{\text{Total time 4 h}} \times 100$			
Lying down: Sleeping	23.0	25.5	11.8
Resting	7.2[a]	12.9[ab]	33.5[b]
Awake	37.2	58.7	45.0
Intermediate	2.7[a]	0.9[b]	1.6[b]
Standing	29.9[a]	2.0[b]	7.9[b]

Groups marked with the same superscript are not statistically different ($P > 0.05$, Mann-Whitney test)

TABLE 4

AMOUNT OF COLOSTRUM INGESTED BY SALERS CALVES 4 AND 10 h AFTER BIRTH

	Sucking the mother	Bottle-fed
Number of calves	21	8
Amount of colostrum (% of the liveweight)	8.9	6.3

The effects of isolation on the development of calf behaviour were observed in heifers of the two breeds, some of which had been isolated and some of which reared with their mother, during the few months before calving and at calving.

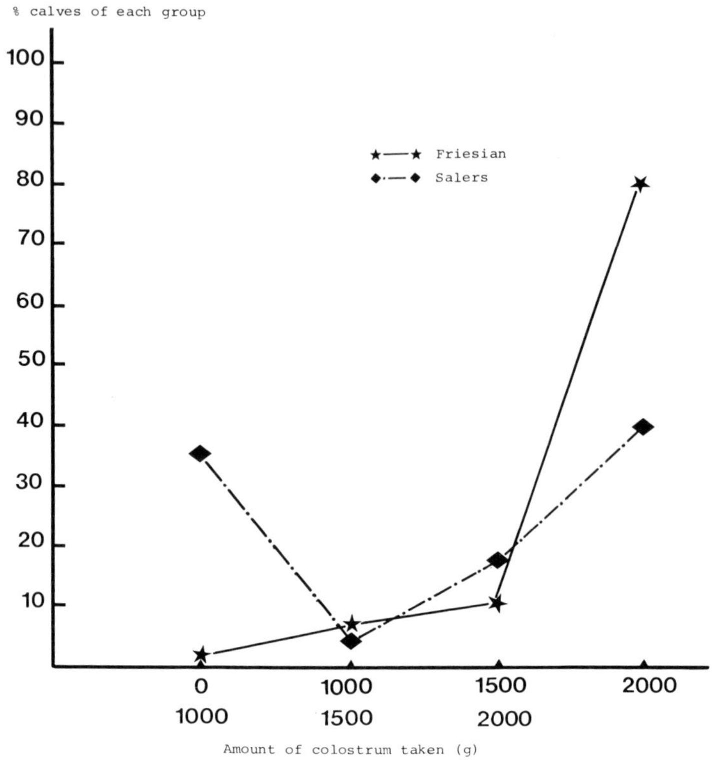

Fig. 5. Intake of colostrum by Friesian and Salers calves.

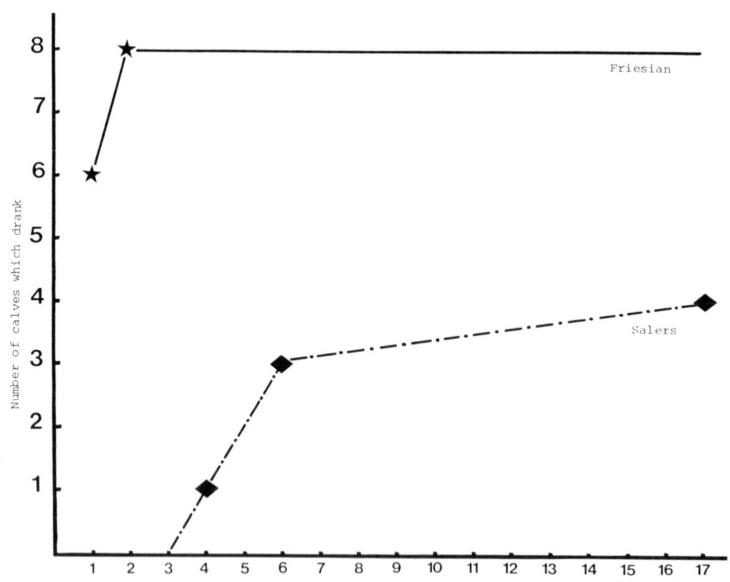

Fig. 6. Adaptation to drinking.

The animals were observed during the 30 h before calving. Table 5 presents the number of interactions between animals which had received the same treatment, combining breed and management system. Salers were more active than Friesians and animals which had been reared with their mothers were more active than those which had been isolated.

TABLE 5

NUMBER OF INTERACTIONS BETWEEN DYADS OF THE SAME GROUP

Relationship		Aggressive	Non-aggressive	Total
Salers: (S)	Mothered (M)	2.28[a]	1.69[ab]	3.97[a]
	Isolated (I)	1.63[b]	1.77[a]	3.40[a]
Friesians: (F)	Mothered (M)	2.03[ab]	0.97[b]	3.00[a]
	Isolated (I)	1.06[b]	0.49[c]	1.55[c]

Groups marked with the same superscript are not statistically different ($P > 0.05$ Kolmogorov-Smirnov test)

The hierarchical power of the animals was calculated in the four herds under observation in which the different treatments were combined. The difference between the breeds was quite clear in that Salers dominated Friesians.

There was no difference between the effects of the management systems for heifers in the Friesian herd, but there were great differences between the two groups of Salers (Table 6). These differences were not as well-defined as they might have been because the isolated Salers heifers were not as heavy as those which had been left with their mothers.

TABLE 6

INFLUENCE OF BREED AND MANAGEMENT SYSTEM DURING EARLY LIFE ON THE HIERARCHICAL POWER* OF HEIFERS

Herd	Friesian (F) Isolated (I)	Friesian (F) Mothered (M)	Salers (S) Isolated (I)	Salers (S) Mothered (M)
SI + SM			4.3 ← ** →	10.6
FI + FM	7.4	6.9		
SM + FM		6.3 ←	** →	10.4
SI + FI	4.7 ←	** →	7.8	

Mann Whitney test $P < 0.05$* $P < 0.01$**

*The hierarchical power is an index of the rank of the animal in the herd (the sum of the animals' hierarchical power in a herd is 100).

No cow displayed abnormal behaviour at calving. The only difference which appeared was the fact that heifers which had been isolated as calves licked their young much more than those which had been reared with their mothers.

CONCLUSION

Although many questions remain unanswered, it is clear that the development of a normal cow-calf relationship ensures the wellbeing of the calf and allows it to become well integrated into the herd. The importance of the mother's past experience, the way in which integration into the herd occurs and in particular the way in which the strong mother-offspring bond ends are still not clear.

Modification of the management system implies adaptation on the part of the animals. It is therefore likely that breeders have selected animals which have been capable of this adaptation. Breeding conditions on a dairy farm, for example, are markedly different from normal conditions. In dairying, breeders have had to select for good activity at birth and good capacity to ingest food, reduced social activity and relatively easy acceptance of isolation in calves, and very low selectivity which allows dairying in cows. The ecological niche built up by the dairy breeder appears to have oriented the evolution of the breed towards a type of animal which presents an extreme adaptability.

It appears, therefore, that to define the welfare of a calf, the breed must first be studied under its usual conditions of management (such as a dairy breed under dairy conditions) before any appreciation of the effects of modifications in the management system can be made.

The results presented here require further confirmation and in particular should be extended to include beef breeds such as Charolais and Limousin, and dairy breeds such as the Holstein. Knowledge of the criteria on which selection for beef or dairying has been carried out will be important, i.e. whether it has been based on multiple characteristics which are each genetically transmitted, or on more general characteristics such as 'emotivity' or 'learning ability'.

ACKNOWLEDGEMENT

I am grateful to Professor D.R. Lindsay for his help in revising the English translation of the manuscript.

REFERENCES

Edwards, S.A. and Broom, D., 1979. The period between birth and first suckling in dairy calves. Res. Vet. Sci., 26, 255-256.

Ewbank, R., 1969. The frequency and duration of the nursing periods in single-suckled Hereford beef cows. Brit. Vet. J., 125 (2), 9-10.

Hafez, E.S.E., 1969. The behaviour of cattle: suckling behaviour. In: The behaviour of domestic animals, London Tindall and Cassell, 253-256.

Le Neindre, P. and Petit, M., 1975. Nombre de tétées et temps de pâturage des veaux dans les troupeaux de vaches allaitantes. Ann. Zootech., 24 (3), 553-558.

Le Neindre, P., Petit, M. and Garel, J.P., 1978. Allaitement de deux veaux par des vaches de race Salers. II. Etude de l'adoption. Ann. Zootech. 27, 553-559.

Le Neindre, P., Menard, M.F. and Garel, J.P., 1979. Suckling and drinking behaviour of new-born calves of beef and dairy cows. Ann. Rech. Vet., 10 (2-3), 211-212.

Nicol, A.M. and Sharafeldin, M.A., 1975. Observations on the behaviour of single-suckled calves from birth to 120 days. Proc. N.Z. Soc. Anim. Prod., 35, 221-230.

Petit, M., 1972. Emploi du temps des troupeaux de vaches-mères et de leurs veaux sur les pâturages d'altitude de l'Aubrac. Ann. Zootech., 21 (1), 5-27.

Poindron, P. and Le Neindre, P., 1975. Comparaison des relations mère-jeune observées lors de la tétée chez la brebis (*Ovis aries*) et chez la vache (*Bos taurus*). Ann. Biol. Anim. Bioch. Biophys., 15, 495-501.

Selman, I.E., McEwan, A.D. and Fisher, E.W., 1970. Studies on natural suckling in cattle during the first eight hours *post partum*. II. Behavioural studies (calves). Anim. Behav., 18, 284-289.

Wagnon, K.A., 1963. Behaviour of beef cows on a Californian range. California Agr. Exp. Stat. Bull., 799, 58 p.

DISCUSSION

F.R. Bell *(UK)* In present systems of dairy farming where the calf is removed at birth or within a few hours is the behaviour of the dam important in terms of evolution or genetics?

P. Le Neindre *(France)* This is certainly a problem and there are great differences in maternal behaviour between beef and dairy cattle. However, this was not the subject of this study although I have studied the differences between dairy and beef calves.

F.R. Bell I was only referring to dairy, but I agree that it is different in the case of beef calves.

P. Le Neindre I think that it is very important that dairy cows have no selectivity, but this is a different point from the behaviour of the calves.

SOME ASPECTS OF THE METABOLIC ADAPTATION OF THE CALF TO AERIAL LIFE

J.F. Grongnet
ENSAR, Zootechnie, 65 rue de Saint-Brieuc F-35042, Rennes, France.

INTRODUCTION

In mammals, birth is not only the beginning of life but must be considered as a stress and an extremely violent metabolic disturbance. In fact, the newborn animal has to take on immediately three vital functions, previously carried out by the dam via the placenta: respiration, thermoregulation and nutrition. Defects in the autonomous development of those functions are the main causes of neonatal death: anoxia, hypothermia, starvation. Among ruminants, lambs are especially sensitive to hypothermia and starvation (Dennis, 1970; Dennis and Nairn, 1970; Dennis, 1972) owing to their small size (Alexander, 1967; Alexander, 1972a and b), the conditions of rearing which are often very rough and the large size of the flocks which makes it impossible to supervise the newborn individually. By contrast the normal lack of such unfavourable factors in most European herds of cattle allows the calf to be protected from such hazards. In the calf, anoxia is the major cause of perinatal death. Prolonged and difficult calvings are frequent. The large size of the foetus which is nearly always single, results in a difficult passage through the mother's genital tract and through the pelvic opening which is often relatively narrow (Philipson, 1976).

In calves, perinatal losses vary from 3 to 30% (Greene, 1978) according to local conditions, breed, management and sanitary conditions in the herd. The economic consequences of such percentages are very important. This is the reason for the development during the last 20 years of research work designed to understand the physiological mechanisms controlling birth: metabolism of the foetus, hormonal control of parturition, causes of distocia, onset of lung function, resistance of the newborn to cold and anoxia, development of the mother-young bond, absorption of immunoglobulins from colostrum, digestive adaptation to milk etc.

It is not realistic to review such a large theme and, consequently, this report will be limited to the presentation of recent results obtained in the INRA research station at Rennes (France). They deal with blood parameters analysed at birth in calves placed in the critical conditions of anoxia, fasting or normal feeding (control).

DEVELOPMENT OF GLYCAEMIA IN THE NEWBORN CALF

Numerous results have been obtained on the postnatal development of glycaemia in the calf but there are discrepancies between different authors. According to Edwards (1969), a continuous increase is observed for the first 24 h, even without any feeding. By contrast, Demigne and Remesy (1979) observed such an increase only after the first meal. Finally, Daniels et al. (1974) and Grongnet et al. (1979) observed a marked increase 30 min after birth followed by a decrease. Furthermore they reported that the influence of the intake of colostrum varied according to when it was given: a limited increase when the first meal was given after 6 h of life, a larger one for the second meal at 12 h and no change for the third one at 18 h.

Such discrepancies have to be carefully considered: individual variations are numerous and results have to be obtained on large numbers of subjects. Furthermore, some changes are of short duration so that very frequent samplings after birth are necessary in order to establish their existence. Finally the circumstances of calving have generally not been carefully considered and reported, nor have the climatic conditions. Our results (Figures 1 - 4), which summarise experiments made for 4 years on 113 subjects, show that all these factors influence the development of glycaemia.

At the very moment of birth, regardless of year, average glycaemia ranges between 0.5 and 0.6 g/l. When calving is in winter and consequently the newborn experiences a relatively cold ambient temperature (from $-8^{\circ}C$ to $5^{\circ}C$) glycaemia increases sharply during the first hour of life to 1.1 - 1.3 g/l. These results from 76 calves over 3 years (Figures 2, 3 and 4) contrast with the absence of any increase in 30 calves born in June of the same years (Figure 1). This supports the assessment of Edwards

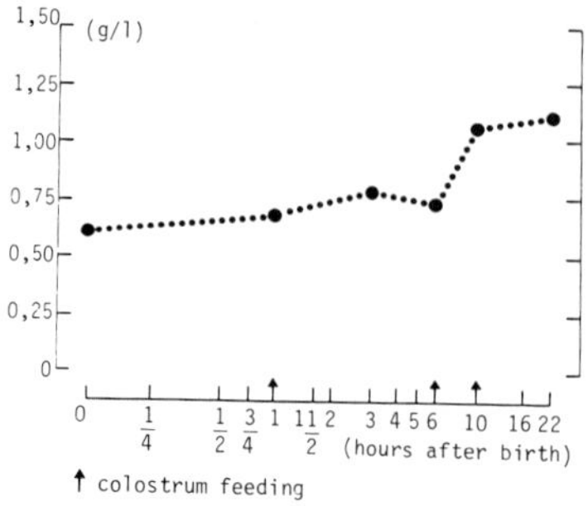

Fig. 1. Evolution of plasma glucose concentration during the neonatal period in 30 dairy calves born in summer and early fed with colostrum.

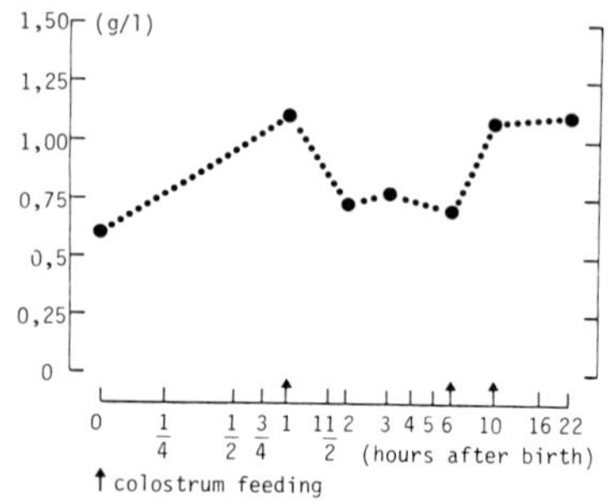

Fig. 2. Evolution of plasma glucose concentration during the neonatal period in 6 dairy calves born in winter and early fed with colostrum.

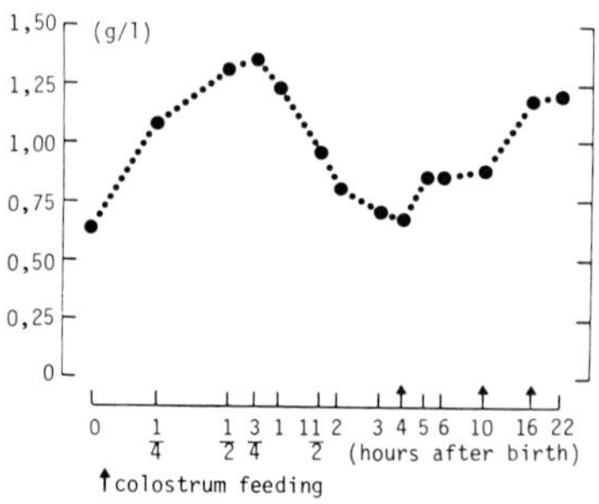

Fig. 3. Evolution of plasma glucose concentration during the neonatal period in 39 dairy calves born in winter and later fed with colostrum.

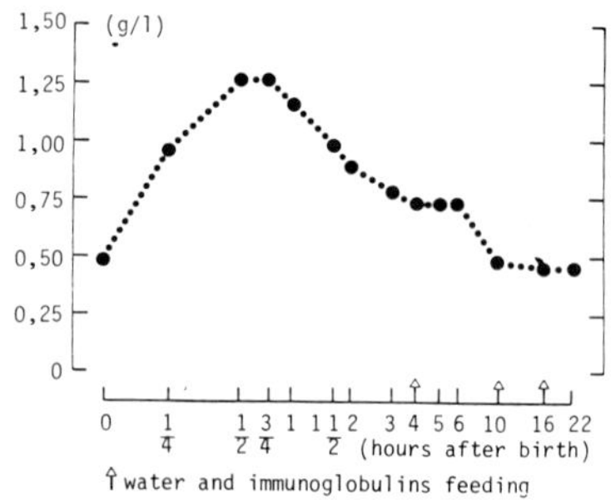

Fig. 4. Evolution of plasma glucose concentration during the neonatal period in 31 dairy calves born in winter and deprived of colostrum.

and Silver (1970) who suggest that the increase in glycaemia during the first hours of life is a consequence of a high sympathetic efferent activity stimulated by cold.

After the first hour of life, the possible increase in glycaemia is followed by a decrease (Figures 2, 3 and 4) even when colostrum is given during this period (Figure 2). This is in agreement with the absence of any increase after colostrum is given in calves born in June (Figure 1). When colostrum is administered for the first time after the fourth hour of life, the increase in glycaemia is very clear (Figure 3) but it should be noted that it rises from a basal level which is already very low (0.5 g/l).

Irrespective of the time of the first drink, a good colostrum supply (1 - 1.5 kg, 3 times) always results in glycaemia values of around 1 g/l at the end of the first day. Conversely, replacing the colostrum with a solution of immunoglobulins which is very low in energy, thus placing the calves in a situation close to complete fasting without exposing the animals to infection hazards, markedly decreases glycaemia. In such a situation, the value is lower at 22 h than at birth. This result does not support the observation of Edwards (1969), which shows an increase over this period even in the absence of any feeding.

To summarise, two successive increases in glycaemia often occur within the first 24 h of the life of the calf. However, they have different origins. The first results from the mobilisation of reserves of glycogen that are high in the newborn. By contrast, the second originates in the food supply and does not occur when there is insufficient food. We can distinguish two mechanisms: the hydrolysis of lactose, although it is relatively scarce in colostrum (Foley et al., 1978), and gluconeogenesis from the amino acids in proteins of the colostrum.

If the reserves of glycogen disappear too rapidly or they are at too low a level, at the beginning, it could lead to marked hypoglycaemia when feeding commences too late. Thus, colostrum feeding has to begin as early as possible. This has the added advantage of assuring good absorption of immunoglobulins. However, even though a satisfactory level of glycaemia is desirable, the absence of an early drink is not too harmful as the newborn is well protected against the adverse effects of hypoglycaemia (Edwards, 1964) by a high level of plasma lactate and by good metabolic utilisation of this molecule (Demigne et al., 1979).

METABOLIC CHARACTERISATION OF ANOXIA IN THE NEWBORN CALF

The work reviewed below was carried out with the aim of developing a means of diagnosis of neonatal anoxia by determining the value of well selected blood parameters. Only the most

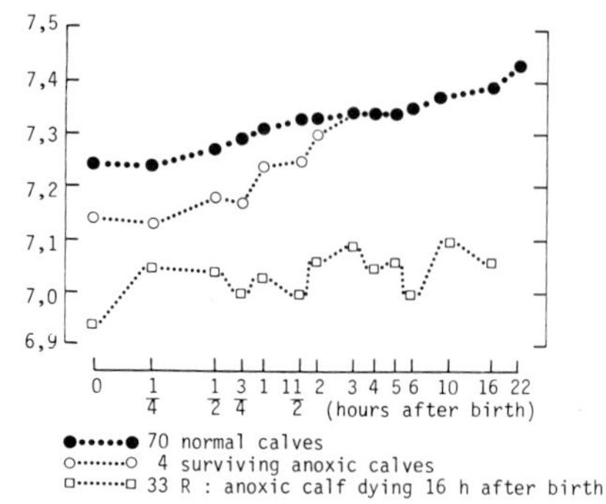

Fig. 5. Evolution of blood pH during the neonatal period in some dairy calves.

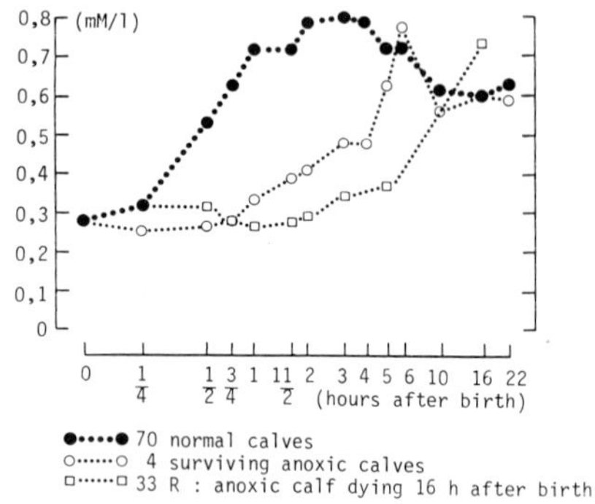

Fig. 6. Evolution of plasma NEFA concentration during the neonatal period in some dairy calves.

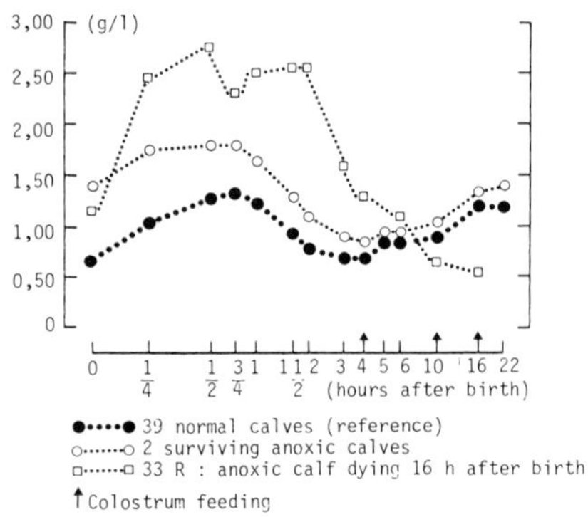

Fig. 7. Evolution of plasma glucose concentration during the neonatal period in some anoxic dairy calves born in winter and colostrum fed.

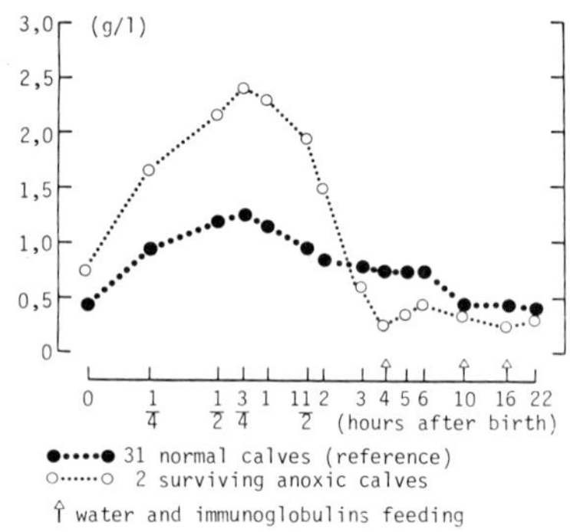

Fig. 8. Evolution of plasma glucose concentration during the neonatal period in some anoxic dairy calves born in winter and colostrum deprived.

significant results will be considered here. Two interdependent lines of research were developed. The first was the study of development of blood criteria immediately after birth in calves born without any trouble either before or during parturition, but of which it could still be claimed that they underwent some kind of subclinical anoxia, as such a phenomenon is always associated with birth. The second is the comparison of the results obtained in normal calves with those from subjects born in a state of marked anoxia. For all the parameters affected by anoxia the graphs for anoxic calves appear to be an exaggeration of the response in normal calves, with the notable exception of the non-esterified fatty acids (NEFA).

The levels are lower for pH (Figure 5), NEFA (Figure 6) and higher for glucose (Figure 7 and 8), lactate (Figure 9), inorganic phosphorus (Figure 10) and uric acid (Figure 11).

pH and lactate

For normal calves, blood pH as measured at the moment of birth is 7.24 (70 subjects) which is in agreement with Massip (1980, 7.26), Maurer-Schweitzer et al. (1977, 7.245), Waizenhofer and Mulling (1978b, 7.209) and Mulling et al. (1972 and 1979, 7.287). In many cases, it decreases slightly during the first 15 min of life as already cited by Mulling et al. (1972) and Maurer-Schweitzer et al. (1977). Later, it increases to reach a plateau 6 to 10 h later, at around 7.35 which is in agreement with the literature.

In anoxic animals, the figures are markedly lower, even below 7.00. The lack of any increase, or sometimes a decrease, is almost certainly a sign of impending death.

The main reason for such an acidosis, limited and transient in normally born calves but more or less marked after anoxia could be associated with the large amount of lactate (Figure 9) which is always present in the blood of newborn calves and is reduced simultaneously with the increase in pH. It results from incomplete glycolysis in the absence of a sufficient amount of oxygen (Savel'eva, 1969). The higher the level of anoxia, the higher are the values. Nevertheless, an intermediate although persistent level observed in calf 33 R which died 16 h *post partum* (Figure 9) could be explained by a total inactivity and a very slow metabolism which is supported by the very low body temperature (from 37°C to 33°C between 2 and 16 h after birth).

Glucose and NEFA

In newborn calves undergoing anoxia, the postnatal increase in glycaemia is always very sudden and markedly higher than in controls (Figures 7 and 8). Such a hyperglycaemic effect of anoxia was demonstrated in the calf by Bloom et al.

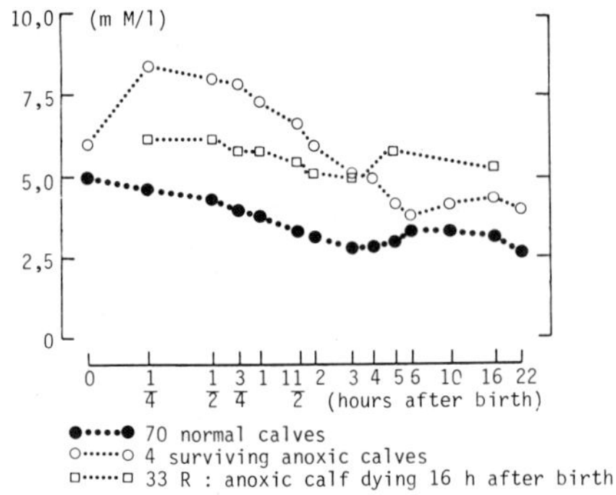

Fig. 9. Evolution of plasma lactate concentration during the neonatal period in some dairy calves.

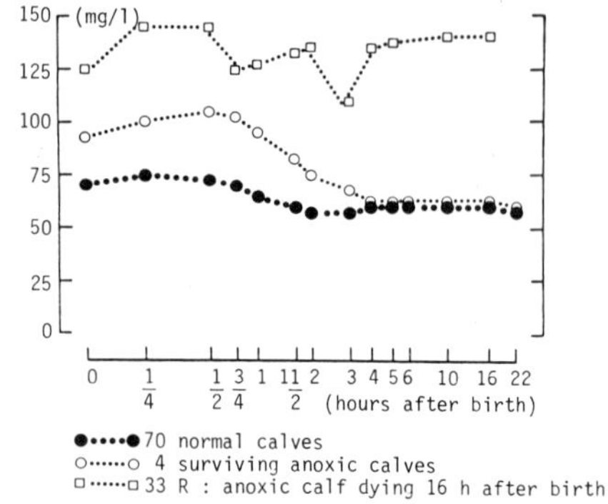

Fig. 10. Evolution of plasma inorganic phosphorus concentration during the neonatal period in some dairy calves.

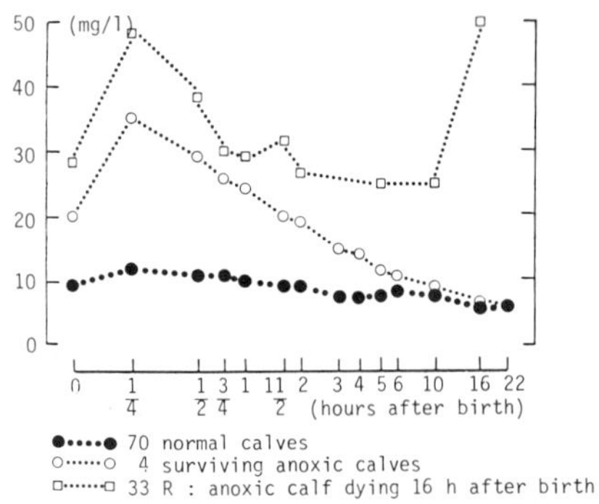

Fig. 11. Evolution of plasma uric acid concentration during the neonatal period in some dairy calves.

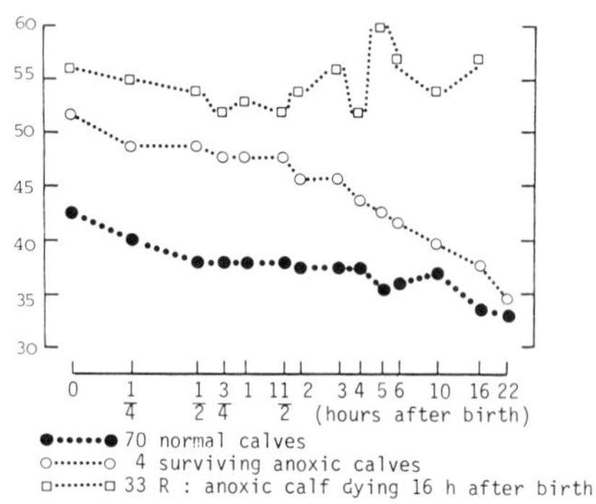

Fig. 12. Evolution of haematocrit during the neonatal period in some dairy calves.

(1976 and 1977) with induced asphyxia and was observed by Kona et al. (1970), at birth. Similar observations were made at birth in newborn puppies (Varma et al., 1973) whereas in the human neonate, anoxia results in hypoglycaemia (Harris et al., 1974).

The cause of such hyperglycaemia in the anoxic calf has not been thoroughly determined as it seems to be very complex. Experimental anoxia induces an increase in circulating cortisol, catecholamines, glucagon and then, as a consequence, insulin. Glucogenolysis is probably considerably increased resulting in a rapid exhaustion of the glycogen reserves. A delay in the first meal may, then, induce a strong hypoglycaemia (0.05 g/l). From the practical point of view, feeding anoxic newborn calves has to be more rapidly done than it is for normal calves.

This hyperglycaemia is always associated with a very low plasma NEFA concentration (Figure 6). This fact has been noted by Harris et al. (1974) in humans and by Kona et al. (1970) in cattle as taking place at the time of birth but has not been explained by these authors. This is a phenomenon which is surprising as the catecholamines and glucagon have lipolytic activity.

Inorganic phosphorus and uric acid

The postnatal patterns of both metabolites have received little attention but their plasma levels decrease during the first hours of life (Figures 10 and 11). The explanation of such a decrease by the disappearance of subclinical anoxia following normal parturition is supported by the fact that very high levels are recorded in calves undergoing anoxia.

In the case of uric acid, inadequate elimination could be due to a temporary malfunction of the kidneys which have a high oxygen demand, whilst in that of phosphorus, with a reduced oxygen supply, the cells would not succeed in reaccumulating ATP from ADP and AMP. An excess of inorganic phosphorus would then diffuse into the plasma and increase phosphoraemia.

Other criteria such as haematocrit, (Figure 12) haemoglobin and creatinine appear to be affected by neonatal anoxia, but the phenomenon itself has to be more clearly established before an explanation should be attempted.

CONCLUSION

Neonatal losses in cattle have an important economic impact in the EC (Mornet et al., 1977). Consequently, an effort has to be made in three areas: at the scientific level, in order to develop a better understanding of neonatal physiology, in the extension service to make methods improving the calf survival better known and at the level of the farmers to develop their understanding of their responsibilities. The sale of large numbers of dairy calves at eight days of age does not stimulate farmers to take an interest in them. This is regrettable because this is a critical period for this weak and slightly mysterious animal, requiring careful management.

A lot of research has been done in the newborn. However, the hours around the time when the young takes its first breath still remain the least understood period in its life. In fact a large number of factors are involved at this moment which are sudden and varied constituting the most important time in the animals life until its death.

REFERENCES

Alexander, G., 1967. Husbandry procedures and neonatal mortality in sheep. Victorian Veterinary proceedings, 70-73.

Alexander, G., 1972a. Neonatal survival in sheep, cattle and pigs: a general approach. Victorian Veterinary proceedings, 70-73.

Alexander, G., 1972b. Survival of the newborn: nutritional factors, environmental and behavioural hazards. Victorian Veterinary proceedings, 73-77.

Bloom, S.R., Edwards, A.V., Hardy, R.N. and Silver, Marian, 1976. Adrenal and pancreatic endocrine responses to hypoxia in the conscious calf. J. Physiol. 261, 271-283.

Bloom, S.R., Edwards, A.V. and Hardy, R.N., 1977. Adrenal and pancreatic endocrine responses to hypoxia and hypercapnia in the calf. J. Physiol., 269, 131-154.

Daniels, L.B., Perkins, J.L., Kriepder, D., Tugwell, M. and Carpenter, D., 1974. Blood glucose and fructose in the newborn ruminant. J. Dairy Sci. 57, 10, 1196-1200.

Demigne, C. and Remesy, C., 1979. Fetal and postnatal metabolism in the calf. Ann. Biol. anim. Bioch. Biophys., 19 (1 B), 159-165.

Demigne, C. and Remesy, C., 1979. Evolution of the postnatal metabolism in the healthy and diarrhoëic calf. Ann. Rech. Vet. 10 (1), 23-31.

Dennis, S.M., 1970. Perinatal lamb mortality in a pure bred Southdown flock. J. Anim. Sci. 31 (1), 76-79.

Dennis, S.M., 1972. Perinatal lamb mortality. Cornell Vet., 62 (2), 253-263.

Dennis, S.M. and Nairn, M.E., 1970. Perinatal lamb mortality in a merino flock in western Australia. Aust. Vet. J., 46 (6), 272-276.

Edwards, A.V., 1964. Resistance to hypoglycemia in the newborn calf. J. Physiol. (London), 171, 46-47 P.

Edwards, A.V., 1969. Carbohydrate metabolism in young animals. In: A.T. Phillipson, Physiology of digestion and metabolism in ruminants, Oriel Press, Newcastle-upon-Tyne, 180-198.

Edwards, A.V. and Silver, Marian, 1976. The glycogenolytic response to stimulation of the splanchnic nerves in adrenalectomised calves. J. Physiol. (London), 211 (1), 109-124.

Foley, J.A. and Otterby, D.E., 1978. Availability, storage, treatment, composition and feeding value of surplus colostrum. A review. J. Dairy Sci., 61, 8, 1033-1060.

Greene, H.J., 1978. Causes of dairy calf mortality. Ir. J. Agric. Res., 17, 295-301.

Grongnet, J.F. and Michel, M.C., 1979. Evolution de quelques paramètres métaboliques chez la vache laitière, au voisinage de la parturition et chez le veau pendant la période néonatale. Sciences Agronomiques Rennes, 79, 170-182.

Harris, J., 1974. Plasma nonesterified fatty acid and blood glucose levels in healthy and hypoxemic newborn infants. J. of Pediatrics 84, 4, 578-584.

Hartmann, H., Mexer, H., Steinach, G., Deschner, F. and Kreutzer, B., 1973. The general adaptation syndrome in calves. I. Normal behaviour of the blood picture and the content of glucose and 11-hydroxycortico-steroids. Arch. Exper. Vet. Med., 27 Heft 5. 811-823.

Kona, E., Kacmarik, J., Sevcik, A. and Elecko, J., 1970. Einfluss der milden Asphyxie neugeborener Kälber auf den Kohlenhydrat und Fettstoffwechsel in den ersten Stunden nach der Geburt. Fortpfl. Haust. Bd. 6 S, 152, 159.

Massip, A., 1980. The relation between the type of delivery and the acid-base and plasma cortisol levels of the newborn calf. Br. Vet. J., 136, 488-491.

Maurer-Schweitzer, H., Wilhelm, U. and Walser, K., 1977. pH values, gas tensions and base excess in the venous blood of normal calves after an uncomplicated delivery. Berl. Münch. Tierärzth. Wschr, 90, 192-196.

Mornet, P. and Quinchon, C., 1977. Epidemiologie. In: Le veau, ed. Mornet, P. and Espinasse, J., Maloine Paris, 231-260.

Mulling, M., Henning, H.J. and Marcks, Ch., 1972. The actual pH values in the blood of newborn calves. Tierärztl Umscham, 27 (4), 180-181.

Mulling, M. Waizenhofer, H. and Brattig, B., 1979. Glucose, Lactate and actual pH values in cows and calves during and immediately after birth. Berl. Münch. Tierärzth Wschr., 92, 111-117.

Phillipson, J., 1976. Studies on calving difficulty, still birth and associated factors in Swedish cattle breeds. Acta Agri. Scand., 26 (2), 151-174 - 26 (3), 211-234.

Savel'eva, G.M., 1969. Use of alkaline solutions in the treatment of asphyxia of the foetus and newborn. Akush. Ginekol., 45 (5), 60-63.

Varma, S., Rakusan, K., Cowan, J.S. and Hetenyi, G., 1973. Stimulation of glucose production by hypoxia in newborn dogs. Can. J. Physiol. pharmacol. 51, 464-471.

Waizenhofer, H. and Mulling, M., 1978a. Lactate, pyruvate and actual pH values in the venous blood of newborn calves. Berl. Münch. Tierärztl. Wschr. 91, 186-188.

Waizenhofer, H. and Mulling, M., 1978b. Research into the behaviour of pH PO_2, PCO_2 in venous, capillary and arterial blood from newborn calves. Berl. Münch. Tierärzth. Wschr., 91, 173-176.

DISCUSSION

F.R. Bell *(UK)* These results are most interesting. How did you know the animals were anoxic?

J.F. Grongnet *(France)* I said that animals were anoxic using classic criteria such as blood pH and lactate. I also said that animals were slightly anoxic because of difficulties during parturition. I wanted to obtain metabolic characterisation so I could not know the result before the experiment. So my aim was to characterise and, using statistical methods, to obtain two populations - normal animals and a well-characterised anoxic population.

F.R. Bell The best way of designating anoxia is to measure the blood oxygen and to measure the blood CO_2. Did you measure these? No, well if you did not measure these then it is possible that the results are due to something else. It is possible for example that the results are due to a hypo-nitraemia.

J.F. Grongnet No, because I have also measured nitraemia which was slightly similar. I measured numerous blood criteria and selected only some for this paper - I measured nitraemia, calcaemia etc. I obtained slight hyper-calcaemia in lungs in strong anoxic conditions but not with my calves. Hyper-calcaemia and not hypo-nitraemia. In our next experiment we have to determine oxygen in arterial blood.

F.R. Bell I was only using hypo-nitraemia as an example that the results might be due to something else than hypoxia, because you really have not measured hypoxia and the results may be due to another reason. I mentioned hypo-nitraemia which you say is not true, but there might be something else.

J.Y. Blom *(Denmark)* How did you take the samples? By an indwelling catheter or did you take a separate sample from the calf every time?

J.F. Grongnet I tried both methods but in my opinion the indwelling catheter stressed the animals more than taking a sample every time.

J.Y. Blom Did you consider the effect of the stress on, for example, glucose levels in the calves?

J.F. Grongnet I believe that calves are not subject to stress by human manipulation during the first few hours after birth. I have sampled animals not using the same technique and the results were similar. It is important for haematocrit because if you sample more and more blood it is possible to have a decrease in the haematocrit. If you sample an animal at birth and 22 h after birth without sampling during the interval the results are the same.

H. Eichinger *(FRG)* How great was the variability of your normal values? Were there significant differences between hypoxia and normal animals? Finally, how can you draw conclusions for animal welfare from these points?

R. Moss *(UK)* Can I ask everyone to remember these three questions because I think it would be wise to leave, particularly the final one, for later discussion. Perhaps we can have just the first two questions answered.

H. Eichinger The first question was how great was the variability? We always have the problem that in special groups of animals, variabilities may be extremely high or very low.

J.F. Grongnet The variability of glycaemia at birth is high but I have presented only significant results. It was impossible to have clear slides showing variability but if you want it is possible to have them, the results really are significant.

H. Eichinger What about the coefficient of variation - 20%, 30% or more?

J.F. Grongnet For glycaemia at birth, the value is near 0.5 plus or minus 0.2 but the minimum and maximum are in a wide scale. The minimum value was about 0.2 and the maximum value was about 1.5, but this is only for special animals.

A REVIEW OF SOME BEHAVIOURAL AND PHYSIOLOGICAL STUDIES WHICH ARE RELEVANT TO THE WELFARE OF YOUNG CALVES

D.B. Stephens

ARC Institute of Animal Physiology,
Babraham, Cambridge CB2 4AT, UK.

ABSTRACT

Most modern intensive husbandry systems which have been specifically designed for the purpose of rearing calves usually involve the calves being fed artificially and housed in groups. The best way of providing optimum conditions to satisfy the behavioural, physiological, hygienic and welfare requirements of such calves has become a matter of concern to the industry.

The practice of using a large 'Continental' beef breed bull on the indigenous dairy herds in Britain has led to a widespread awareness of the problems associated with parturition of the calves from such matings. If the parturition is protracted, the calf is often lacking in vigour. This lack of vigour can lead to a prolongation of the time taken by the calf to locate a teat and suck colostrum thus lowering the neonatal calf's chances of survival. Other factors under genetic control, such as the anatomical characteristics of the udder of the cow, can also be of critical importance. If the udder is too large, then the calf is often unable to locate the teat and suck within the first few hours of birth.

Many of the methods currently employed for the artificial rearing of calves involve the calves being housed in groups. Usually this leads naturally to the setting up of a social hierarchy, often within the first few hours of the group being established. A disadvantage of such systems is that abnormal oral and ingestive behaviour patterns frequently occur and can have significant hygienic implications. For example, ingestion of hair or non-food material often results in digestive upsets and sometimes there is preputial cross-sucking. If left unattended, such behaviour patterns can eventually prove fatal for the affected calf, usually due to starvation. Experiments in our laboratory have provided results which suggest that the dopaminergic neuronal system is implicated in such stereotyped abnormal oral behaviour.

The influence of various modern husbandry practices on cardiovascular and neurohormonal parameters has been the subject of several recent reports. In our own laboratory, changes in heart rate, blood pressure and renal blood flow were detected in calves when exposed to various normal routine husbandry practices such as restraining for venipuncture and transportation.

The relationship between the physical environment and the behavioural requirements for hygiene of calves has also been examined experimentally in our own and other laboratories. For example, decreasing the living space, as often happens under conditions of veal housing, was found to produce only a small and transient increase in heart rate and plasma corticosteroid concentration levels. In another experiment, it was shown that calves prefer light to darkness.

It was concluded on the basis of numerous observations involving both field and laboratory situations, conditions which were assumed to be adverse for the animal did in fact produce physiological and behavioural changes which were detectable in the calf. In addition, the degree of response was graded, usually to an extent which reflected the intensity of the stimulus. At a practical level, it is probably best to assess the requirements of the calf within each specific situation. For example, provision should be made, where necessary, for at least a proportion of each day to be spent with the calf house illuminated.

INTRODUCTION

Systems of husbandry which were initially developed for the poultry and pig industry are now being employed in calf rearing systems. Such intensive husbandry techniques often involve the animals being grouped and kept in increasingly large numbers on each unit. The disturbances produced by movement to such units while still very young are likely to influence the behaviour and development of the calf. Both the welfare implications and the production considerations which are associated with all aspects of intensive methods of calf rearing have become a matter of some concern to the industry. Considerations of the problems associated with relating animal welfare to intensive methods of animal production have been reviewed by Brambell, 1965 and Murphy, 1978. The scientific methods available for assessment of welfare and stress in farm animals have been reviewed by Duncan, 1974, 1975 and Stephens, 1980.

In many respects, the behavioural repertoire of the young calf is less complete and probably less competent than that of the adult. However, each developmental stage of the calf's behaviour has evolved so that it can cope with the succession of new problems that it encounters from the moment of birth. Learning, which enables behaviour to be progressively modified in the light of successive experiences, probably goes on throughout the entire life of the animal. Therefore, it is often difficult to recognise and delineate precisely when each stage in behavioural and physiological development has been reached. A parallel development in the organisation of the physiological capabilities of the calf is also essential to allow this ontogeny of behaviour which enables the calf to adapt to the progressively changing requirements of the environment.

The aim of this review is to present and evaluate information on the behavioural and physiological responses of calves reared under both field (i.e. natural or normal husbandry) and laboratory conditions.

THE INFLUENCE OF BREEDING POLICY ON THE VIABILITY OF THE NEONATAL CALF

The breeds of parents (dam and sire) and feeding practices employed for the dam can greatly influence the viability of the neonatal calf. There are several factors that can be of importance. Among these are:-

Parturition effects on the vigour of the neonatal calf

It is clear that many of the factors that can influence behaviour of the calf are innately present and are therefore under genetic control.

The vigour and speed with which a calf becomes active after birth can often depend on the ease with which parturition takes place. The level of activity of both mother and neonate in the period immediately after birth can have an important bearing on the degree of mutual contact and therefore the establishment of the bond between mother and young. For example, skin contact between woman and newborn baby has been shown to be a major factor in establishing such a bond in the human (Jelliffe and Jelliffe, 1978). The ease with which parturition occurs in the bovine depends upon several factors. For example, changes in economic conditions in agriculture, particularly as regards the methods of feeding beef cattle, often lead to changes in breeding practices (Glücksmann, 1981). Breeding programmes have been devised which ensure that certain features, such as the ability of specific sires to transmit the genetic potential for rapid growth rates to their hybrid progeny.

This practice of using a larger terminal sire on the smaller dairy herds can, however, result in complications at parturition, which can depress the vigour and therefore the viability of the calf. The size of the newborn calf, in particular its anatomical dimensions in relation to the birth canal of the dam, is dependent on several factors. Firstly, birth size can be critically dependent on the nutritional status of the dam at parturition. This clearly depends on the level of feeding during the last 6 - 8 weeks of pregnancy. For example, dairy cows are often fed extra amounts near term ('steaming up') to increase milk yield. Overfeeding leads to an increase in size of the calf which can result in a difficult parturition. Gestation length, which is partly breed dependent, is also important. For example, the larger breeds, such as the Charolais, have a gestation length of between 285 - 295 days when purebred. (This compares with the average of around 279 - 280 days that is normally quoted for the more common British dairy breeds such as Jerseys and Friesians when purebred). Even when, as is commonly practised commercially, a smaller breed of dam, such as the Jersey or Friesian is bred to a larger terminal sire such as the Charolais, Simmental, Blonde d'Aquitaine, Chianina, Maine Anjou or Limousin, the gestation length is often extended to between 285 - 290 days. Since rapid growth of the foetus occurs during the final stages of gestation, a prolonged gestation results in a bigger calf at birth, and therefore the possibility of serious dystokia is very greatly enhanced. The concomitant deleterious effects on the vigour, learning ability, particularly in relation to sucking, can then be of vital importance for the viability of the calf unless skilful stockmanship is applied to ensure early sucking of colostrum.

Sucking behaviour

The locating of a teat, which is the first of several interrelated phases making up the sucking response (Baliassnikowa and Model, 1931) is of critical importance to a mammal. Although the initial approaches of the neonatal mammal towards its dam are stereotyped (Sluckin, 1972), it is probable that learning plays an important role in the sucking response and in the establishment of an enduring bond with its dam. It was shown in our laboratory that the development and nature of sucking behaviour of newborn goats contains a significant learning component (Stephens and Linzell, 1974). In this study, parturition and behaviour of mother and neonate were observed in goats in which one of the two mammary glands of the dam had been transplanted to the neck. In this situation, the newborn kid was offered the choice of moving anteriorly or posteriorly in order to locate a teat. It was found that movement of the kid was directed forwards and backwards randomly along the mobile rounded form of the dam's ventral surfaces until the smooth, hairless surface of the mammary gland and teat was located, and then sucking would begin. Skin contact between woman and newborn baby has also been shown to be an important component of the sucking response in the human (Jelliffe and Jelliffe, 1978).

Relation between the anatomy of the udder, sucking behaviour and the calf's requirements for hygiene

In the same study (Stephens and Linzell, 1974), it was noted that when the mammary gland of the goat was large and pendulous, considerably more time was taken for the kid to locate a teat and start sucking. Unfortunately, similar considerations have very often been found to apply in cases of calves born to some of the modern dairy breeds, whose udders are often large and oedematous at the time of parturition. Under such conditions, the time taken for the calf to locate the teat and to suck the vital colostrum can be greatly extended (Edwards, 1981). Therefore, careful genetic selection of cows with the appropriate anatomical conformation of the udder can greatly enhance the viability of the neonatal calf. Fortunately, such characteristics of the udder which are highly desirable from the point of view of enhancing the chances of the calf sucking very soon after birth, are also important for ease of applying the machine for milking and so are given serious consideration during genetic selection procedures.

In practice, however, the udder is very often large, pendulous and perhaps with large teats located in awkward positions so that sucking by the calf shortly after birth is less likely. If such conditions are suspected, it is essential that colostrum be obtained artificially from the dam (or alternatively thawed from a prepared colostrum 'bank') and then offered as soon as possible to the newborn calf via a rubber teat. It has been suggested that a minimum of two litres of colostrum should be ingested as soon after birth as circumstances allow and preferably under sterile conditions (Brand, 1981, personal communication).

Immunological protection

It has been reported that up to 50% of all young calves under 1 week of age have a sub-optimal level of immunity due to insufficient intake of colostrum (Edwards, 1981).

The importance of the mother-young bond

Antibodies from the blood of mammals appear in the colostrum. It has also been shown (Jelliffe and Jelliffe, 1978) that in the woman, specific antibodies are produced locally by the reticulo-endothelial tissue of the breast and are discharged into the milk within 8 h of immunological stimulation. If such a mechanism of rapid diathelic immunisation occurs in the bovine, it would provide an important mechanism of immunological protection which was specific for each mother-infant pair. This means that the process of sucking in the period immediately surrounding parturition, antigens from the calf and its environment, could rapidly stimulate specific antibodies in the dam which would then be transferred via the colostrum to the calf. These antibodies would become immediately available to give protection appropriate for the environment. Such antibodies would be in addition to those in the colostrum derived from the mothers established immune system.

Requirements for hygiene

The conditions of hygiene under which the calf learns to suck are also important. In the 'trial and error' learning process, where the calf searches randomly along the under surface of the dam, any loose hair and surface dirt is likely to be licked and ingested. Such material can mix with the mucus material lining the surface of the villi in the intestine to form a coating thus creating a mechanical barrier which greatly impairs the rate and efficiency with which the immunoglobulins in the colostrum are able to penetrate the crypts between the villi. Clearly, access to the whole surface area available at the villi greatly facilitates the rapid absorption of immunoglobulins (Brand, 1981, personal communication).

EXAMPLES OF VARIOUS METHODS EMPLOYED FOR THE ARTIFICIAL REARING OF CALVES

The husbandry practices used for rearing calves vary enormously. In Britain, the majority of calves reared for beef are produced from dairy cows. Those that are surplus to requirements as replacements for the breeding herd and which also possess an adequate conformation are usually reared for beef production. The age at which they are removed from their natural dam and taken for rearing normally varies between 1 and 3 weeks. The usual methods of rearing that are employed include:

1) The multiple suckling of calves in a series of small groups, usually 4 or less at any one time, sequentially on a foster mother. In this way it is usually possible for 20 calves to be reared per lactation;

2) The use of artificial milk substitutes at body temperature directly from buckets on a once or twice daily basis. Sometimes the buckets are fitted with a teat and valve arrangement to aid those calves who appear unable to learn to drink directly;

3) Milk provided via a teat from an apparatus designed and programmed to dispense freshly mixed milk continuously at body temperature, usually on an *ad libitum* basis;

4) A very recent development, whereby milk substitute products are acidified, usually with propionic acid, to give a pH of around 4 in the reconstituted milk.

The development and use of acidified milk replacers has meant that reconstituted milk can be stored for several days without souring, thus increasing the potential for group rearing. It also has the labour saving advantage of allowing mechanisation of the mixing process, on a twice weekly basis for example, and the milk can then be stored safely in bulk quantities at ambient temperature. Such 'cold' milk is acceptable to calves and is made available on an *ad libitum* basis to the group from several rubber teats, which are fixed to the pen at head height for the calf, and connected via tubes to a large centralised communal milk container. A non-return valve system is incorporated at one end of each connecting tube so that the system is fully primed at all times and so ensures that milk is constantly present within each teat. The constant presence of milk within the teat tends to encourage the sucking process. This is important, especially in winter, when the milk may be very cold and less readily acceptable.

THE INFLUENCE OF METHOD OF REARING ON THE BEHAVIOUR AND HYGIENE OF THE CALF

Single suckled calves, left to suck their dams naturally, form a close bond with her only. Other calves in the herd are usually of a different age and therefore at a different stage of development. For artificial rearing however, it is usually endeavoured to match calves in groups to be as near to each other in age and size as is possible. This has important behavioural implications.

Social behaviour of groups

These various 'group' systems of calf rearing therefore normally depend on calves being reared in batches, which naturally leads to the setting up of a hierarchial social organisation. The development and organisation of the social behaviour seen in groups of calves reared under one such artificial system has been described (Stephens, 1972; 1974). In these studies following a road journey of about 8 h, British Friesian male calves aged between 1 and 3 weeks were matched for size and placed in groups (usually 15 in number). Each calf was taught on an individual basis to suck from a rubber teat which provided milk at body temperature from a milk dispensing machine. Usually, by 3 or 4 days after arrival, the majority of the calves were able to locate the teat and proceed to suck on their own. At this stage, it was also evident that a fairly stable social hierarchy had become established. The dominant calves, which were usually the larger ones with a body weight in excess of 50 kg, were rarely interrupted and displaced from the teat while sucking. In contrast, a relatively smaller, subordinate calf in the group, was often interrupted during a bout of sucking by a dominant member of the group which approached it and displaced it from the teat. This displacement of subordinate calves from the teat during sucking resulted in a marked reduction in the total sucking times of such calves. However, to compensate, the subordinate calves spent more time at the dry feed troughs and therefore consumed more dry feed and hay. Although the dominant calves grew at a faster rate while milk was available (usually 6 weeks), this advantage was not maintained in the immediate post weaning period. This was probably because the subordinate calves, which were more conditioned to eating dry food, suffered less of a nutritive setback as a consequence of the change produced by the abrupt removal of the milk dispensing machine. However, 3 weeks after weaning, the dominant calves had regained their weight advantage over the smaller subordinate members of the group.

Similar rank orders in groups of calves, based upon competitive agonistic interactions, have also been described by Broom and Lever (1978). These authors found that the established social grouping remained stable when the same groups of calves were kept together for as long as 8 to 20 months. The rapid tendency of calves to form an organised social unit when put together into groups has also been described by Kilgour et al., (1975).

Abnormal oral and ingestive behaviour

Field observations

In newborn mammals, the sucking response is one of their earliest and most complex behavioural patterns. Therefore, it is not surprising that when young animals, such as calves, are reared artificially, abnormal tendencies in their ingestive behaviour often develop.

Various forms of abnormal ingestive behaviour and cross sucking in calves have been described by several authors (Hafez and Lineweaver, 1968; Stephens, 1972, 1974; Wood et al., 1967). Calves will often lick and chew the wooden hay racks and sometimes the wall or other objects in the house. One of the major undesirable features observed where calves are kept in groups is that the calves are provided with ample opportunities for cross sucking. Examples of appendages crossed-sucked were ears, scrotum and in particular the prepuce (Stephens, 1972, 1974). One calf was observed to suck its own prepuce and release urine which was ingested. Evidence that the sucking drive is highly complex and is not entirely satisfied by an artificial rubber teat was provided in a study in piglets (Stephens, 1975). Using a device whereby the voluntary milk intake could be monitored independently from the oral activity, it was shown that the animal persisted in vigorous oral activity, even when the nutritive requirements had been satisfied by pre-loading the stomach with milk directly via an intragastric tube. These results in the piglets supported the conclusions of Levy (1934) and Ross (1951) that a sucking need exists in the very young mammal which is not entirely directed towards satisfaction of the animal's nutritive requirements. Evidence for the oral-deprivation hypothesis as a major factor for the aetiology of cross-sucking was provided in a survey reported by Wood et al., (1967). They found that the incidence of cross-sucking in calves reared on their dams was only about half of that of calves fed milk directly from a bucket. These authors further suggested that the most effective method for controlling the cross-sucking was to ensure an adequate supply of concentrate feed was available *ad libitum* and also to delay the mixing of calves into groups until they were about a month of age. Such abnormal behavioural developments did not seem related to any apparent dietary deficiencies.

Laboratory experiments

Experiments in our laboratory have used techniques in an attempt to provide a possible neurochemical basis to explain the development of abnormal oral behaviour in farm animals (Sharman and Stephens, 1975; Fry et al., 1975). It was noted (Stephens, 1972, 1974) that when an abnormal oral behaviour (in particular preputial cross-sucking) developed in a particular calf in the group, then the abnormal behaviour invariably became escalated in its intensity. The behaviour seemed to be compulsive and virtually impossible to eradicate and such calves would eventually have to be isolated from the group. The abnormal oral behaviours observed in calves appeared to be stereotyped in nature, and bore a striking similarity to the stereotyped behaviours seen in laboratory rodents given injections of apomorphine. Although this drug produces emesis in the human and dog, its effect in other species such as the sheep and calf (Sharman and Stephens, 1974) is to produce a characteristic stereotyped oral behaviour. There is evidence that some of the behavioural effects seen as a result of administration of apomorphine are due, at least in part, to the stimulation of dopamine receptors at sites in the brain such as the corpus striatum (Ernst and Smelik, 1966; Ernst, 1967; Anden et al., 1967). When apomorphine was administered intravenously to a calf there were immediate pronounced licking movements which accompanied lateral side to side movements of the head. Following this stereotyped oral behaviour there was active licking of the straw bedding and the edge of the water bowl. These observations would therefore appear to implicate the dopaminergic neuronal system in the development and mediation of the onset of abnormal oral behaviour seen in calves reared artificially.

THE INFLUENCE OF MODERN HUSBANDRY PRACTICES ON SOME PHYSIOLOGICAL PARAMETERS OF EMOTIONAL RESPONSES IN CALVES

Routine handling

Calves are frequently handled at different stages of their lives for such purposes as identification marking, transportation, marketing, therapeutic disease control programmes. In such circumstances, behavioural, psychological and physiological responses are elicited which provide for the maintenance of homeostasis. In particular, the hypothalamus-pituitary-adrenal system is normally activated under such conditions. The general nature of the physiological responses that are stimulated and the sequence of integrated events involving the nervous and endocrine systems following central perception of an aversive stimulus have been outlined in reviews on laboratory animals (Gray, 1971) and, more recently, on farm animals (Stephens, 1980). In the rat, for example, it has been shown that the extent of the changes in concentration of plasma corticosteroids can sensitively reflect varying intensities of psychological stimulation (Hennessy and Levine, 1978).

Due to the ease of recording fluctuations in heart rate (HR) directly and also by telemetry, this parameter has been used extensively to study the responses of calves when exposed to emotional disturbances. The significance of the HR changes which accompany any adverse stimuli have been reviewed in detail by Sutterer and Obrist (1972). Changes in renal blood flow (RBF) also occur in response to emotional arousal (Henry et al., 1973).

Experiments have been carried out with Jersey calves at our Institute to monitor simultaneously:- (a) cardiovascular responses and (b) the accompanying changes in plasma concentrations of corticosteroids which result from some modern routine husbandry practices. Manual restraint to allow blood samples to be taken from the jugular vein produced an elevation in HR from the resting level of 100 beats per minute (b.p.m.) to over 140 b.p.m. A ten minute restraint period produced a quadrupling in plasma corticosteroid concentrations. Physical hand restraint, however, produced only a small decrease in RBF. By contrast, the RBF was reduced to below 50% of the resting level in one young calf for the whole period during which it sucked its milk allowance via a rubber teat (Figure 1). When sucking, calves frequently close their eyes and become relatively oblivious to their environment. Using an anthropomorphic analogy, an observer would describe the sucking process by a hungry calf as 'a pleasurable sensation'. The evidence that there is a marked renal vasoconstriction during sucking clearly points to the fact that teat sucking can produce an acute emotional arousal in the calf.

In another study (Stephens and Toner, 1975) it was found that if the experimenter (who was familiar to the calf) entered the loose box and simply walked around the calf, even without touching the animal, the HR increased from 90 to 125 b.p.m. and the HR remained at this level for 2 - 3 min. The calf showed no overt signs of being disturbed, other than to appear mildly aroused and turning its head to take an 'investigative look' at the experimenter. The sight of food elevated the HR to 135 b.p.m. When eating dry food, the rate lowered to 108 b.p.m. These observations emphasise the importance of the need for

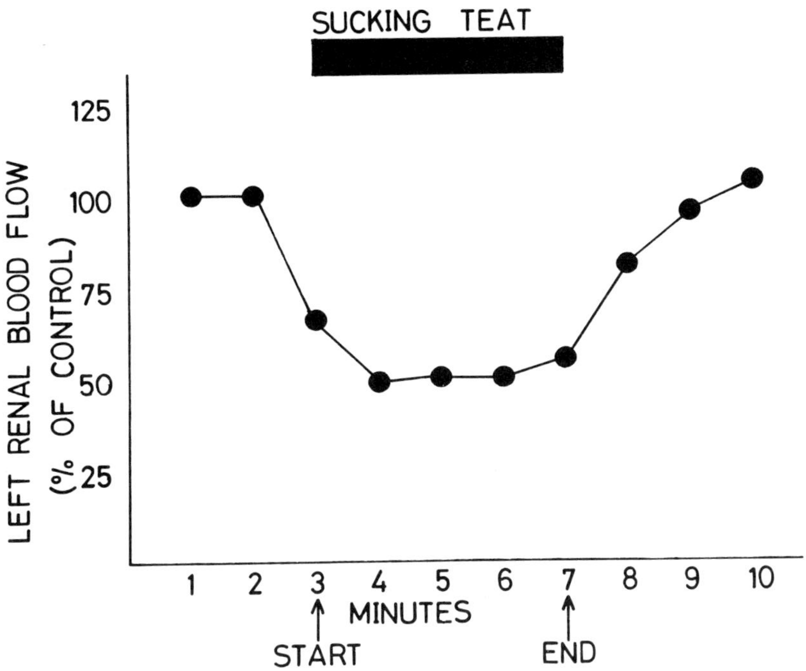

Fig. 1. Change in renal blood flow as a percentage of control in a calf before, during and after being allowed to suck its milk from a teat.

considering the environmental conditions as well as the previous experiences of the calf when interpreting physiological parameters which are monitored. In the sheep, for example, it has been shown that a surprisingly prolonged period was required for adaptation to laboratory conditions (Pearson and Mellor, 1976). A minimum period of 6 weeks was required before the sheep could be handled with a little change in HR. Changing climatic conditions produce changes in the metabolic rate of young calves. It has been found (Holmes et al., 1976) that the HR also varies with climatic change and hence HR fluctuations provided a fairly reliable indicator of variations in metabolic rate.

It is now well established that, in laboratory animals such as rats, the pituitary-adrenal hormones have pronounced behavioural effects (de Weid et al., 1975; de Weid, 1977; Van Wimersma Greidanus, 1970, 1977). Similar observations have recently been made in pigs by Mormède and Dantzer (1978).

The behavioural changes resulting from the fluctuations in blood concentration of adrenal hormones released due to disturbances such as those described above in the calf under various husbandry conditions do not appear to have been studied.

Transportation

Geographical variations between regions usually result in different crops being husbanded in each region. Therefore, it often happens that areas where food suitable for fattening calves is grown (usually grain) are located long distances away from areas where food suitable for dairying (usually grass) is grown. These facts require that calves are transported, often over quite long distances. The physiological range of responses resulting from such road journeys and the extent to which the calves can cope, assume practical importance. It has been shown in New Zealand (Kilgour and Mullord, 1973) that if calves are transported over very long distances, then every consideration should be given to practices aimed at minimising the disturbance to the calves during the first day in particular of a journey, since the patterns of travel and adaptive responses are established during the early stages. Johnston and Buckland (1976), found that transportation lasting as little as 1 - 4 h produced a larger elevation of plasma corticosteroids than routine procedures such as debudding or castration. Calves aged between 1 and 4 weeks (Kent, 1977), were less severely distressed by one 18 h journey than by 14 h of travel interspersed with periods of rest which prolonged the journey time to span 2 days. This indicates that the routine pauses in normal journeys are not adequate to enable the calf to recover fully or adapt.

Observations on the physiological responses of calves to transportation in our own laboratory have been described in detail in another volume in this series. There was a greater increase in HR and plasma concentration of corticosteroids when the calves were transported in a free standing position, than when the same journey was repeated with the calf kept within a wooden cubicle in the lorry. When the calves laid down, however, there was a significant decrease in the HR compared with when they were standing (Stephens and Toner, 1974, 1975; Toner and Stephens, 1975). It is not clear why the heart rate was reduced when the calf became recumbent. Any redistribution of blood between organs which accompanies changes in posture may result in a reflex decrease in HR. In the human (Dauncey and James, 1979) there was a similar effect, where it was shown that there was a marked increase in HR when the subject stood up. A factor which may have contributed to the decrease in HR when the calves laid down was an accompanying decrease in the level of emotional arousal.

PHYSICAL ENVIRONMENT IN RELATION TO THE BEHAVIOURAL AND WELFARE REQUIREMENTS OF CALVES

Clearly, the total environment into which the calf is born and survives is complex. Bond et al. (1964) defined environment as 'all conditions, circumstances and influences surrounding and affecting the development of an organism or groups ... often contrasted with heredity'. The thermal relationship of the neonate with its environment is very important (Mount, 1968). This is particularly so in hairless animals such as the piglet (Stephens, 1972) and also calves if housed in sub-standard conditions which are wet and draughty. For the purposes of this review, the following aspects of the influence of the physical environment on the calf have been selected and are briefly discussed.

Space requirements

The amount of floor space which is optimal to satisfy the requirements of calves kept under intensive husbandry conditions is a subject of some concern to the calf industry, both from the points of view of hygiene and welfare. The requirement for space by calves in relation to the whole environment, and the size and age of the group have been widely discussed. The first major document on this subject was the Brambell Report (1965). This was followed by Codes of Recommendation for the Welfare of Livestock, of the Ministry of Agriculture, Fisheries and Food (1971) and also later discussed in a symposium organised and reported by Universities Federation Animal Welfare (1978). A common feature of all these reports is that they considered it essential that each calf should have enough space to enable it to turn and groom itself.

In experiments on Jersey calves in our Institute (Stephens and Toner, 1974, 1975; Toner and Stephens, 1975) it was found that moving a calf from a normal loose box into a type of wooden crate similar to that often employed for veal production (i.e. significantly reducing the living space of the calf) produced only a transient increase in HR and the levels of plasma corticosteroid concentration in the period immediately following the change. During the ensuing 14 day period, while the calf remained in the cubicle, the HR remained near the normal pre-trial control levels. These observations suggest that physiological parameters such as HR are not necessarily accurate indices of the effects of prolonged restriction of living space. Clearly, more experimental work, particularly observational studies, need to be done to establish the space requirements which allow calves to achieve their optimal behavioural needs and thereby maintain their well-being.

Illumination levels

It has recently been shown (Baldwin and Start, 1981) that calves will perform operant responses in order to obtain artificial light when housed in darkened, light proof chambers. It was found that in light-dark preference tests, 67% of each 24 h period was spent in the light thus providing objective evidence that calves prefer light to darkness.

Climatic conditions

The effects of both short and long term changes in ambient temperature and the concomitant physiological responses to climatic changes in the bovine has been reviewed in detail by Thompson (1973). The first responses to a sudden change in environmental temperature, such as panting in the heat or shivering in the cold, begin within minutes and reach a steady level within hours. Exposure of calves to minus (-) 4°C within a few hours after birth does not produce a sustained elevation in plasma corticosteroids (Khan et al., 1970). It would appear desirable, from the point of view of energy conservation, that excessive heat loss should be prevented by providing suitable housing accommodation. A 'wind-chill index' has been developed for cattle to describe the effects on heat loss of modifying factors such as radiation exchange by providing various forms of shelters (Webster, 1971).

Design of handling facilities

The design of handling facilities for calves when being moved in groups for handling or transport purposes onto vehicles, has been reviewed by Grandin (1978). It was shown, for example, that animals will follow a curved path more readily than a straight path. Calves also seem to prefer to move up a gentle slope, and will often hesitate when forced to move down an incline. Observations also show that calves prefer to be moved as part of a semi-cohesive mass or better still be allowed to remain within an established social group. Moving calves in groups where possible, has further advantages. Firstly, it makes the task easier for the personnel responsible for the calves. Secondly, there is probably less stimulation of the hypothalamus-pituitary-adrenal system. This may have an important bearing on the young calves' ability to: (a) learn and habituate to the novel environment (de Weid, 1977) and (b) to cope with the bacterial and viral challenge resulting from handling and moving to new environments. The increased release of corticosteroids into the plasma as a result of exposure to any form of adverse situation may have an important effect on the total immunity possessed by the calf. For example, administration of corticosteroids to calves soon after birth has been shown (Husband et al., 1973) to halve the efficiency of absorption of colostral immunoglobins. In the same study, it was also shown that endogenous production of some types of immunoglobins was delayed by administration of corticosteroids.

CONCLUDING REMARKS

From the studies described, it seems to be established that in the calf, those situations which were assumed to be adverse for the animal do in fact produce physiological and behavioural changes which are detectable and that the degree of response induced reflects the intensity of the stimulus. One problem that seems difficult to interpret is the observation that the cardiovascular and endocrine changes which are normally assumed to provide a profile of the response to stress may also occur during conditions which are usually assumed to be 'pleasurable'. For example, the sympathetic stimulation during sucking which affects the blood flow in the renal artery in the calf. Therefore, at present, there appears to be no immediate prospect of a universal test which can be applied to evaluate the precise requirements of the calf in all situations. At a practical level, we will have to content ourselves with attempting to assess the behavioural, physiological and requirements for hygiene of the calf within each specific situation. For example, light should be provided for at least part of each day. We must also be careful to try to consider all the variables which are operating in each husbandry situation.

It must be realised that the results obtained in laboratory experiments, where usually one or two parameters are measured at any one time under controlled conditions, may not always be directly applicable to the more complex field situation. A combination of ethological observations under husbandry conditions plus physiological and behavioural studies in the laboratory is essential in order to evaluate welfare problems in farm animals.

ACKNOWLEDGEMENTS

I wish to thank Dr. A. Glücksmann, of this Institute, for help in preparing the manuscript, Messrs. Stubbs, Ellern Mede, Totteridge Common, London N20 8LS, for their constructive comments and also Miss C.E. Delaney for skilful technical assistance.

REFERENCES

Anden, N.B., Rubenson, A., Fuxe, F. and Hokfelt, T., 1967. Evidence for dopamine receptor stimulation by apomorphine. J. Pharm. Pharmac. 19, 627-629.

Baldwin, B.A. and Start, I.B., 1981. Sensory reinforcement and illumination preference in sheep and calves. Proc. R. Soc. Lond. B211, 513-526.

Baliassnikowa, N.J. and Model, H.H., 1931. Zur Neurologie des Saugens. Such. f. Kinderforsch. 39, 1-16.

Bond, T.E., Heitman, Jr., M. and Kelly, C.F., 1964. Physiological response time of thermally stressed swine to several cooling media. VIth Internat. Contr. Agr. Engrg. Lausan 1, 356-370.

Brambell, F.W.R., 1965. 'Report of the Technical Committee to enquire into the Welfare of Animals kept under Intensive Livestock Husbandry Systems'. Cmnd. 2836, London, HMSO.

Brand, A., 1981. Personal communication. Clinic of Veterinary Obstetrics, Gynaecology and AI, University of Utrecht, Yalelaan 7, De Uithof, Utrecht, Netherlands.

Broom, D.M. and Leaver, J.D., 1978. Effects of group-rearing or partial isolation on later social behaviour of calves. Anim. Behav. 26, 1255-1263.

Codes of Recommendations for the Welfare of Livestock, 1971. Code No. 1 Cattle. Min. of Agric. Fish. and Food.

Dauncey, M.J. and James, W.P.T., 1979. Assessment of the heart-rate method for determining energy expenditure in man, using a whole-body calorimeter. Br. J. Nutr. 42, 1-13.

Duncan, I.J.H., 1974. A scientific assessment of welfare. Proc. Soc. Anim. Prod. 3, 9-19.

Duncan, I.J.H., 1975. Research methods in farm animal welfare. Study Tour Report (1975 Winston Churchill Memorial Trust Fellow). Poultry Research Centre, Edinburgh.

Edwards, S.A., 1981. Factors affecting the latency to suckling in dairy calves. Paper No. 1. Proc. of British Society of Animal Production. Winter Meeting 23-25 March 1981. Anim. Prod. (in press).

Ernst, A.M., 1967. The mode of action of apomorphine and dexamphetamine on gnawing compulsion in rats. Psychopharmacologia (Berl.) 10, 316-323.

Ernst, A.M. and Shelik, P.G., 1966. Site of action of dopamine and apomorphine on compulsive gnawing behaviour in rats. Experientia 22, 837-838.

Fry, J.P., Sharman, D.F and Stephens, D.B., 1975. A possible neurochemical basis for abnormal behaviour in farm animals. Proc. of International Symposium on Catecholamines and Stress. Eds. E. Usdin, R. Kvetnansky and I. Kopin, p. 603. Pergamon Press.

Glücksmann, A., 1981. Sexual dimorphism in human and mammalian biology and pathology. Academic Press, London.

Grandin, T., 1978. Design of lairage, yard and race systems for handling cattle in abbatoirs, auctions, ranches, restraining chutes and dipping vats. In: 1st World Congress on Ethology Applied to Zootechnics. Industrias Gráficas España, S.L. Comandante Zorita, 48-Madrid-20, p. 37-52.

Gray, J.A., 1971. The psychology of fear and stress. Weidenfeld and Nicolson.

Hafez, E.S.E. and Lineweaver, J.A., 1968. Suckling behaviour in natural and artificially fed neonate calves. Z. Tierpsychol. 25, 187-198.

Hennessy, M.B. and Levine, S., 1978. Sensitive pituitary-adrenal responsiveness to varying intensities of psychological stimulation. Physiol. Behav. 21, 295-297.

Henry, J.P., Meehan, J.P., Rader, R.D. and Stevens, C.M., 1973. Renal vasoconstriction following psychosocial stimulation. Bull. Los Angeles Co. Med. Assoc. November 1st.

Holmes, C.W., Stephens, D.B. and Toner, J.N., 1976. The measurement of heart rate as an indicator of the energy metabolism of calves kept out-of-doors. Livestock Production Science 3, 333-341.

Husband, A.J., Brandon, M.R. and Lascelles, A.K., 1973. The effect of corticosteroid on absorption and endogenous production of immunoglobulins in calves. Ajebak 51, 707-710.

Jelliffe, D.B. and Jelliffe, E.F.P., 1978. Human milk in the modern world. Oxford University Press.

Johnston, J.D. and Buckland, R.B., 1976. Response of male Holstein calves from seven sires to four management stresses as measured by plasma corticoid levels. Can. J. Anim. Sci. 56, 727-732.

Kent, J.E., 1977. The effect of road transportation on young cattle. M.Sc. Thesis, The University of Liverpool.

Khan, M.A., Dickson, W.M. and Meyers, K.M., 1970. The effect of low environmental temperature on plasma corticosteroid and glucose concentrations in the newborn calf. J. Endocr. 48, 355-363.

Kilgour, R. and Mullord, M., 1973. Transport of calves by road. N.Z. Vet. J. 21, 7-10.

Kilgour, R. Winfield, C.G., Bremner, K.J., Mullord, M.M., De Langen, H. and Hudson, S.J., 1975. Behaviour of early-weaned calves in indoor individual cubicles and group pens. N.Z. Vet. J. 23, 119-123.

Levy, D.M., 1934. Experiments on the suckling reflex and social behaviour of sheep. Am. J. Orthopsychiat. 4, 203-224.

Mormède, P. and Dantzer, R., 1978. Pituitary-adrenal influences on avoidance behaviour of pigs. Hormones and Behaviour 10, 285-297.

Mount, L.E., 1968. The Climatic Physiology of the Pig. Edward Arnold.

Murphy, L.B., 1978. A review of animal welfare and intensive animal production. Poultry section, Queensland Department of Primary Industries, Australia.

Pearson, R.A. and Mellor, D.J., 1976. Some behavioural and physiological changes in pregnant goats and sheep during adaptation to laboratory conditions. Res. Vet. Sci. 20, 215-217.

Ross, S., 1951. Suckling behaviour in neonate dogs. J. abnorm. soc. Psychol. 46, 142-149.

Sharman, D.F. and Stephens, D.B., 1974. The effect of apomorphine on the behaviour of farm animals. J. Physiol. 242, 25-27P.

Sluckin, W., 1972. Imprinting and early learning. Manuals of Modern Psychology, Methuen.

Staples, G.E. and Haugse, C.N., 1974. Losses in young calves after transportation. Br. vet. J. 130, 374-379.

Stephens, D.B., 1972. Piglet survival: a review of some physiological considerations. Vet. Annual: 64-73. John Wright and Sons Ltd., Bristol.

Stephens, D.B., 1972. Social behaviour in groups of artificially reared bull calves. Proc. World Assoc. for Buiatrics Congress, London, pp. 476-480.

Stephens, D.B., 1974. Studies on the effect of social environment on the behaviour and growth rates of artificially-reared British Friesian male calves. Anim. Prod. 18, 23-34.

Stephens, D.B., 1975. The effects of gastric loading on the sucking response and voluntary milk intake in neonatal piglets. J. comp. physiol. Psychol. 88, 2, 796-805.

Stephens, D.B., 1980. Stress and its measurements in domestic animals: a review of behavioural and physiological studies under field and laboratory situations. Adv. vet. sci. comp. med. 24, 179-210.

Stephens, D.B. and Linzell, J.L., 1974. The development of sucking behaviour in the newborn goat. Anim. Behav. 22, 628-633.

Stephens, D.B. and Toner, J.N., 1974. A method for the continuous display of heart rate in the calf during transportation. J. Physiol. 242, 24-25P.

Stephens, D.B. and Toner, J.N., 1975. Husbandry influences on some physiological parameters of emotional responses in calves. Appl. Anim. Ethol. 1, 233-243.

Sutterer, J.R. and Obrist, P.A., 1972. Heart rate and general activity alterations of dogs during several aversive conditioning procedures. J. comp. physiol. Psychol. 80, 2, 314-326.

The Welfare of the Food Animals 1979. Proc. of a symposium organised by The Universities Federation for Animal Welfare, 8 Hamilton Close, South Mimms, Potters Bar, Hertfordshire.

Thompson, G.E., 1973. Review of the progress of dairy science climatic physiology of cattle. J. Dairy Res. 40, 441-473.

Toner, J.N. and Stephens, D.B., 1975. Instantaneous heart rate indicator with digital display. Laboratory Practice, July 1975, 474.

Webster, A.J.F., 1971. Prediction of heat losses from cattle exposed to cold outdoor environments. J. Appl. Physiol. 30, 5, 684-690.

De Weid, D., 1977. Peptides and behaviour. Life Sci. 20, 195-204.

De Weid, D., Witter, A. and Greven, H.M., 1975. Behaviourally active ACTH analogues. Biochem. Pharmacol. 24, 1463-1468.

Van Wimersma Greidanus, Tj.B., 1970. Effects of steroids on extinction of an avoidance response in rats. A structure-activity relationship study. In De. de Weid and J.A.W.M. Weijnen (Eds.), Progress in Brain Research, Vol. 32, pp. 185-191. Elsevier, Amsterdam.

Van Wimersma Greidanus, Tj.B., 1977. Pregnene-type steroids and impairment of passive avoidance behaviour in rats. Horm. Behav. 9, 49-56.

Wood, P.D.P., Smith, G.F. and Lisle, M.F., 1967. A survey of inter-sucking in dairy herds in England and Wales. Vet. Rec. 81, 396-398.

DISCUSSION

L.H. Huisman *(Netherlands)* Do you have any solution for avoiding urine sucking?

D.B. Stephens *(UK)* We tried putting evil-tasting material onto the prepuce of the calves, we also tried putting it onto the nose of the sucking calf, we tried reteaching the calf, but if nothing seemed to work then they had to be removed.

N. Steenkamer *(Netherlands)* Dr. Stephens, you showed in a diagram that with the help of blockers you could reverse the effect of apomorphine.

D.B. Stephens Yes, we could block the effect of apomorphine.

N. Steenkamer Did you work with these blockers in order to avoid the navel sucking and other bad habits? Theoretically, it seems to be a good idea and might be a practical solution as well, but up till now no real answer has been given.

D.B. Stephens All we have done so far is to put in blockers before putting in the apomorphine, so blocking the apomorphine effects. Your question was that if we had abnormal sucking animals, would blockers solve the problem? The answer is that it works for a few hours, but as soon as the tranquillising effect has passed off due to metabolism the problem unfortunately returns.

The guinea-pig, cheap to keep, does this scrotal sucking and we can use this as our model.

H. Eichinger *(FRG)* You said that 15 calves form a social group. Is this simply due to the capacity of the artificial nurse, or are there other reasons?

D.B. Stephens It is due in part to the capacity of the nurset. Also the calves were kept in old converted stables and we judged from the space available that a good group would consist of 15 animals. There was nothing magical about the number.

H.C. Adler *(Denmark)* I would like to discuss with Dr. Stephens a problem that has arisen in recent years. We have, of course, all the various types of sucking found in weaning calves that you have mentioned but in addition we have the problem of adult cows sucking other cows. Have you any explanation for this? Do you think it is a consequence of deprived calf suckling? Does this mean that the desire to suck will never leave the animal?

D.B. Stephens I have no explanations but I can describe my experiences. It happens in the Jersey cow and seems to be seasonal during spring and summer Sometimes the problem disappears.

H.C. Adler I agree that it occurs mostly in Jerseys but also occasionally in other breeds e.g. Red Danish. I would say that it occurs all the year round rather than on a seasonal basis and I have not known the problem to stop, only to get worse.

D.B. Stephens I meant that it becomes infectious throughout the herd, in forty cows you can get 20 doing this cross-sucking. These are my experiences from observing one or two Jersey herds.

We suggested that they should be given a molasses ball - as the cow licks the ball, it turns and picks up fresh molasses. We found that this helped in this particular case and the following year the problem did not persist.

G. van Putten *(Netherlands)* Dr. Stephens asked the maybe theoretical question about what hormone levels and biochemical results mean, and to this you could well add physiological results. I think it should be made clear that these results are simply an aid to help us to quantify ethological observations.

J.P. Signoret *(France)* My first comment is about this question of cross-sucking among adult cows. I obtained figures from people in the DDR that this was markedly influenced by genetics and we could try to relate this genetic capacity not only for sucking other cows but also of tolerating being suckled. Both are important. The recipient cow accepts something that is absolutely abnormal maternal behaviour. This could relate to the possible selection made against maternal behaviour and against the strength of the maternal drive that occurs when selecting for milk production. We select for the amount of milk produced and, as Le Neindre said in his paper, this is a selection for an easier discharge of oxytocin etc., not controlled by the brain in relation to the maternal behaviour. So there could be some relation between this type of milk selection - independent of the brain mechanism in relation to maternal behaviour - and the appearance of such troubles that occur more and more often, the more intensive is the selection for milk production.

Another comment concerns the development of such cross-sucking of ears, teats, etc. in lambs and calves. Have you observed any correlation between drinking from a bucket or actively sucking? In our experience with lambs, when the teats are hard to suckle and when the milk is available on demand, then there is practically no cross-sucking with lambs. What happens with your nurset compared with twice-a-day bucket drinking? Is cross-sucking less or more frequent?

D.B. Stephens I have not looked at this in relation to calves but with lambs we were able to go in and make some experimental changes. We tried putting a rubber tube above the teat so that the head could be touched, we tried to ensure that the valve systems worked so that the teats were always primed and I should add that the milk was cold. We were unable, however, to eradicate this in the lambs despite everything that we tried. We also tried putting divisions between the teats so that the lambs would not touch each other and so trigger it off. It seemed to us that whatever we did, once the lambs started the cross-sucking we could not eradicate it. You said that if you have warm milk available *ad libitum* you can get rid of it?

J.P. Signoret This cross-sucking is as you have described it in calves. But with lambs, in the artificial feeding situation, the cross-sucking is not more frequent than when they are suckled by the ewes. I do not know why.

J.Y. Blom *(Denmark)* I would like to draw your attention to some recent Norwegian research in this area. Cross-sucking is very interesting and I think the solution to it should not be medication but rather knowledge of ethology. In this Norwegian study, they put calves in boxes, four in each box, and put one suckling bucket with teats for each calf in the boxes. Milk was put into these every morning and every night but there emerged a clear pattern of behaviour. The calves went to the bucket every fourth hour to suck even if there was no milk in it. This arrangement immediately stopped cross-sucking so perhaps this could be a solution.

BEHAVIOURAL AND PHYSIOLOGICAL STUDIES ON REARING CALVES AND VEAL CALVES

J. Unshelm, U. Andreae and D. Smidt
Institut für Tierzucht und Tierverhalten, Mariensee der FAL, D-3057 Neustadt 1, Federal Republic of Germany.

As a result of various observations, questions have arisen as to whether procedures used for fattening veal calves and rearing calves have an adverse effect on the wellbeing of calves. This paper deals with some problems relating to welfare, mainly of veal calves, which have become apparent in the course of multifaceted research projects which have been carried out during the last few years.

The following topics will be briefly discussed:

- the haemoglobin concentration in the blood;
- the development of trichobezoars in the rumen;
- the influence of floor conditions on the hooves;
- the occurrence of ulcers in the abomasum;
- changes in behaviour related to housing and management.

The problem of low haemoglobin concentrations in the blood of veal calves arises from iron-deficiency anaemia caused by feeding regimes designed to obtain the 'white' veal favoured by consumers. Two groups of calves kept in single boxes were compared in order to investigate this phenomenon under different feeding and management conditions. One group of black and white calves was kept on slatted floors and the other in straw boxes. Both groups were fed milk replacer *ad libitum*, whilst the group kept on straw had access to this additional source of iron. Each animal in this group consumed 200 g/day wheat straw.

Table 1 shows the erythrocyte count, the haemoglobin concentration, the haematocrit, the iron concentration and the iron binding capacity of the calves at the age of six months. The figures indicate that the majority of values for the group kept on straw were significantly better than those of the group kept on the slatted floors.

Table 2 shows the individual haemoglobin concentrations of all the animals between the ages of one and six months.

TABLE 1

THE RED BLOOD PICTURE AND IRON CONCENTRATION IN 6 MONTH OLD VEAL CALVES KEPT ON SLATTED FLOORS OR IN STRAW BOXES

Parameter	Slatted floor (n = 13) $\bar{x}$	s	Straw (n = 14) $\bar{x}$	s	Difference
Erythrocytes (Mill/µl)	5.18	1.15	5.07	0.63	0.11
Haemoglobin (g/100 ml)	7.5	1.2	9.3	1.2	1.8**
Haematocrit (%)	23.6	3.2	27.5	2.8	3.9**
Iron concentration (µg/100 ml)	83.0	20.0	181.0	39.0	98.0**
Iron binding capacity (µg/100 ml)	490.0	3.0	476.0	23.0	15.0*

* $p \leqslant 0.05$; ** $p \leqslant 0.01$

Values below 6 g/100 ml (underlined) occurred only in the group which was kept on the slatted floor and which had no access to straw. Abnormal behavioural symptoms such as staggering when walking to the scale and excessively prolonged lying periods were observed in these animals. Both these ethological criteria are indicated in Table 2 by asterisks and are exclusive to the animals in the group deprived of straw shown in the lower half of the Table. Two animals with haemoglobin levels below 5 g/ 100 ml died. It would seem from these values that a minimum of 6 g/100 ml is too low from the point of view of the welfare of the animals (Unshelm, 1980).

Giving access to eating straw is not the only way of solving iron deficiency problems and solutions can be found in other management practices. Furthermore, the additional iron supplied by straw might be dependent on varying iron contents of different sources of straw. The intake of a certain amount of 'structured feed', however, does seem to be of considerable importance in preventing the formation of so-called hairballs (trichobezoars) in calves fed exclusively on milk replacers (Bogner, 1978; Unshelm et al., 1979; Groth et al., 1979).

TABLE 2

INDIVIDUAL HAEMOGLOBIN VALUES IN VEAL CALVES KEPT ON SLATTED FLOORS OR IN STRAW BOXES

Animal no.	Floor	Haemoglobin concentration g/100 ml at various ages 1 month	2 months	3 months	4 months	5 months	6 months
01	straw	9.6	9.4	10.9	10.2	10.3	10.8
02	"	8.0	7.4	7.1	7.4	7.4	7.6
03	"	7.4	6.7	8.4	8.4	9.0	8.8
04	"	13.4	11.8	10.0	9.9	10.3	10.3
05	"	12.0	10.3	9.5	9.4	9.4	10.3
06	"	8.2	9.5	10.4	10.3	10.4	10.6
07	"	10.2	8.7	9.4	10.2	9.1	7.9
11	"	10.3	10.1	8.6	9.2	9.2	9.3
12	"	8.0	7.3	7.6	8.3	8.7	9.6
13	"	10.6	10.5	8.5	8.1	7.8	8.5
14	"	9.1	10.2	10.5	10.4	10.1	8.8
15	"	13.4	10.3	8.6	9.3	9.9	10.7
16	"	14.1	12.4	10.3	9.8	8.7	9.8
17	"	7.7	7.5	6.8	8.1	7.2	7.3
* 51	slats	12.7	11.8	9.6	8.4	7.8	7.2
* 52	"	9.8	9.4	7.7	6.1	6.0	5.4
53	"	8.1	7.2	6.0	6.3	7.4	7.0
54	"	11.8	11.1	12.0	10.7	9.8	9.4
55	"	11.1	10.7	8.5	8.0	8.1	7.9
56	"	12.7	11.6	9.5	8.8	9.2	8.7
* 57	"	9.0	8.7	6.7	6.8	6.7	5.7
* 58	"	7.0	6.5	4.9	5.3	5.3	+
* 61	"	12.7	12.5	8.1	9.2	8.8	8.6
62	"	11.5	9.5	8.3	8.2	8.5	8.8
* 63	"	5.0	3.9	4.6	3.6	+	+
* 64	"	8.4	8.0	8.1	7.1	6.8	7.4
65	"	10.6	9.2	9.8	8.2	7.6	7.7
66	"	9.4	8.4	7.9	6.0	6.3	7.3
* 67	"	7.1	5.9	6.7	5.7	6.8	6.7

Underlined values = Hb concentration below (<) 6 g/100 ml
6 g/100 ml = 60 g/l = 3.72 mmol/l

* Animals with increased susceptibility as, for instance, indicated by a staggering way of walking on the way to the scale

\+ Animal died during the experiment

In the studies carried out on this problem the formation of trichobezoars could only be avoided by giving a daily ration of 200 g of structured feed (straw, chopped straw, straw cobs) from the third week of age. If straw was not given until the calves were five weeks old this resulted in the formation of hairballs in one in four of the animals. If straw was not given until the calves were thirteen weeks old, all of them were found to have bezoars. The number of bezoars varied from 1 to 27 per calf, and the size of bezoar varied from that of a tennis ball to 1 kg in weight. The effect of structured feed in preventing bezoar formation is not that the animals cease to lick hair, but that the passage of ingested hair through the digestive tract is secured.

Intake of roughage in order to prevent bezoar formation is not, of course, dependent on the use of straw as a bedding material. The general statement that calves are likely to feel more comfortable on straw than they are on slatted floors cannot be rejected, from the point of view of welfare, in relation to the use of slatted floors in veal calf boxes. Certain findings which may be relevant to calf welfare do not, however, support the idea that slatted floors are inferior. When comparing the incidence of pathological symptoms in the hooves, such as microlesions and acute and chronic haemorrhages, Brentano et al. (1979) showed that keeping calves on straw had a negative effect. The above symptoms were found in 72% of animals kept on straw and in only 23% of the calves kept in boxes with slatted floors. This may be due to the softer and more succulent condition of the hooves of those calves kept on straw.

The occurrence of abomasal ulcers in fattening calves is still an unsolved aetiological problem. In our observations, irrespective of other management factors, 90% of calves with unlimited access to milk or milk replacers showed abomasal ulcers. Ulceration occurred even in suckler calves, regardless of whether or not roughage was provided

It may tentatively be concluded from these findings that the occurrence of abomasal ulcers is not necessarily connected either with stress induced by inappropriate handling or with the ingestion of roughage. However, the ulcers could well be related to feeding intensity, as they can be avoided by decreasing the daily amount and concentration of milk replacer.

Table 3 shows some results from our survey of behavioural characteristics. These results indicate that the ingestion of approximately 200 g/day of straw improves rumination and reduces the frequency of behavioural substitutes such as gnawing on wood and licking objects. So-called 'tongue-playing' has been widely interpreted as being a consequence of suckling deprivation. This assumption must be contested since tongue-playing is occasionally also observed in suckling calves (nursing on cows or on artificial teats) and the frequency of occurrence of such behaviour can be greatly reduced by offering roughage. Self-licking which, as mentioned previously, causes bezoar formation when roughage is not offered in the diet, can be seen in both suckling calves and calves drinking from buckets and is therefore not a sign of behavioural deviation. However, mutual sucking, mostly on specific regions of the partner's body, is a problem when calves are housed together and bucket fed. These substitutional activities usually begin straight after feeding and are only performed for a short period of time (Table 4). They did not give rise to umbilical inflammation in the animals observed. Some calves did, however, develop the habit of drinking urine and seemed to favour urine over milk replacer. These animals had to be transferred to individual boxes (Zeeb et al., 1970; Scheuermann, 1974; Graf et al., 1976; Papendiek, 1979; Andreae, 1980; Andreae et al., 1980).

CONCLUSIONS

The rearing and fattening of calves to final weights of 200 kg or more using only milk replacers as feed may result in welfare problems such as anaemia, formation of trichobezoars and behavioural abnormalities. These can largely be prevented by giving the animal access to straw or other structured feed, although there are other methods of preventing severe anaemia. Access to straw does not necessarily involve the use of straw as a bedding material, since the prevalence of hoof lesions decreased considerably when calves were kept on slatted floors

TABLE 3

BEHAVIOURAL ACTIVITIES* IN 12 WEEK OLD VEAL CALVES (± ROUGHAGE, DRINKING OR SUCKLING SYSTEM)

Housing		Deep litter			Slatted floor	Indiv. H	Group H	Deep litter
Supplementary feed		--"--	------	Straw cobs	------	Chopped straw		hay + cob centr. fed
		n = 14	n = 14	n = 14	n = 15	n = 17	n = 16	n = 7
Behavioural characteristics								
Ruminating	$\bar{x}$	51.2	[12.0]	50.3	[21.8]	29.8	38.3	
	s	(12.9)	(7.7)	(11.1)	(8.8)	(15.7)	(11.6)	
Gnawing at wood	$\bar{x}$	2.4	[8.1]	3.4	4.6	1.4	1.2	1.5
	s	(1.5)	(4.3)	(1.8)	(2.3)	(1.2)	(0.9)	(0.8)
Licking objects	$\bar{x}$	1.7	4.0	10.2	12.0	14.6	10.7	4.9
	s	(1.5)	(3.0)	(3.3)	(4.0)	(4.1)	(2.7)	(1.8)
Licking themselves	$\bar{x}$	11.5	8.3	11.8	10.7	16.3	22.3	14.1
	s	(3.6)	(4.4)	(4.8)	(3.4)	(4.4)	(12.2)	(2.7)
Tongue playing	$\bar{x}$	0.1	1.3	2.7	[5.5]	1.6	0.9	0.2
	s	(0.4)	(1.7)	(1.5)	(5.8)	(1.0)	(0.5)	(0.3)

* Tested in 5 minute intervals, 4 x 24 h of observation.

[ ] distinct frequency of vacuum/substitute activities produced by lack of roughage.

TABLE 4

THE FREQUENCY OF SUBSTITUTE SUCKLING BETWEEN VEAL CALVES DRINKING FROM ARTIFICIAL TEATS OR BUCKETS RESPECTIVELY

Behavioural activities		Artificial teat n = 12	Bucket n = 12
Suckling on praeputium	$\bar{x}$	±0	3.0
	s	-	(5.2)
	min/max	-	0 - 16.5
Suckling on scrotum	$\bar{x}$	0.7	5.0
	s	(0.9)	(5.8)
	min/max	0 - 2.5	0 - 18.5
Suckling on coat	$\bar{x}$	0.2	1.0
	s	(0.3)	(2.9)
	min/max	0 - 0.5	0 - 10

rather than on bedding material. This survey was based on simultaneously recorded ethological, physiological, biochemical and pathological criteria. Interdisciplinary investigation therefore seems to be an important prerequisite for a reasonable approach to animal welfare problems (Smidt et al., 1980).

REFERENCES

Andreae, U., 1980. Verhaltenskriterien als tierschutzrelvante Indikatoren bei Mastbullen und Mastkälbern. Landbauforschung Völkenrode, Sonderheft 53, 67-73.

Andreae, U., Unshelm, J. and Smidt, D., 1980. Handhabung von Kälbern in Gruppenhaltung. In: Aktuelle Arbeiten zur artgemässen Tierhaltung 1979. KTBL-Schrift 254, 89-96.

Bogner, H., 1978. Einige Aspekte zur Haltung und Fütterung von Kälbern aus der Sicht des Tierschutzes. Berl. Münch. Tierärztl. Wschr. 91, 459-463.

Brentano, G., Dämmrich, K. and Unshelm, J., 1979. Untersuchungen über Gelenk- und Klauenveränderungen bei auf Lattenrosten und auf Stroheinstreu gehaltenen Mastkälbern. Berl. Münch. Tierärztl. Wschr. 92, 229-233.

Graf, B., Wegmann, R. and Rist, M. 1976. Das Verhalten von Mastkälbern bei verschiedenen Haltungsformen. Schweizerische Landwirtschaftliche Monatshefte 54, 333-335.

Papendiek, T., 1979. Ethologische Reaktionen von Kälbern auf Rohfasermangel bei Haltung auf Lattenrosten. Landbauforschung Völkenrode, Sonderheft 48, 62-66.

Smidt, D., Andreae, U. and Unshelm, J., 1980. Ist "Wohlbefinden" messbar? Anmerkungen zu einem Tierschutzproblem. Tierzüchter 32, 338-340.

Unshelm, J., 1980. Verhaltenphysiologische Indikatoren für tierschutzgerechte Haltung von Mastbullen und Mastkälbern. Landbauforschung Völkenrode, Sonderheft 53, 74-85.

Unshelm, J., Andrea, U. and Smidt, D., 1979. Verhaltensphysiologische Studien an Mastkälbern und deren Bedeutung für die Kälberhaltung. Tierzüchter 31, 499-500.

DISCUSSION

A.J. Webster *(UK)* What was the iron content of your milk replacer diet?

D.T. Smidt *(FRG)* I am afraid I cannot answer that question. It was commercial milk replacer but I do not have the exact value.

N. Steenkamer *(Netherlands)* These calves were probably fed for the first six or seven weeks on a starter milk replacer containing between 50 - 60 ppm iron and later on a finisher diet containing some 10 ppm iron.

A.J. Webster The question that emerges from that, and perhaps Dr. Steenkamer may like to answer it, is why do you think straw removed the problem of iron deficiency anaemia when the amount of iron present in 200 g of straw/day would be negligible?

D.T. Smidt The straw is not solely responsible for the change in the haemoglobin levels etc. but we believe that it is also necessary in calves of such a high weight to develop the rumen and to prevent hairball formation.

A.J. Webster I would say that in our experience we have identical blood haemoglobin levels with veal calves in crates and on straw, but in certain circumstances when veal calves are raised on straw in groups the blood haemoglobin can be higher because the food conversion ratio is lower, and therefore the animals grow more slowly on the same amount of energy. Iron is an accumulated nutrient and so in these circumstances they can have a higher haemoglobin.

N. Steenkamer This discussion is very interesting but I think that the main reason why calves fed straw in addition to milk replacer in some cases are less anaemic and in other cases are as anaemic as the control calves, is simply because the iron content of straw varies greatly. You can have a sample of barley straw containing some 15 ppm iron and so if the calves consume 200 g or 1 kg/day the haemoglobin level will not be affected. However, you can have samples of barley straw with 450 ppm iron and 200 g consumption of this per day will really affect haemoglobin levels as well as meat colour. I think the main difference in the report on this effect of straw lies in the iron content, perhaps copper too, of the straw.

P.J. O'Connor *(Ireland)* What sort of straw was used - barley, wheat or oats?

D.T. Smidt I do not know all the details but I believe it was wheat straw.

G. van Putten *(Netherlands)* I would like to ask about the abomasal ulcers - a subject that has interested us for the last year or two. We inspected about 1 000 abomasums of veal calves, half of which were given straw and the other half only milk substitute. There were considerably more ulcers or lesions in those given additional straw, the difference was about 20%. I agree with you that straw is not the cause of these ulcers or lesions but I do think straw aggravates the problem, maybe because it causes a lower pH in the abomasum because of the operation of the rumen.

D.T. Smidt We also thought that this might be the case but when we looked we found few real differences and indeed found these ulcers in nearly 100% of calves suckled by the cow. We concluded from this that straw feeding was not to blame for the formation of ulcers.

G. van Putten How many animals were involved in this observation?

D.T. Smidt Not as many as 1000, but a large enough number upon which to draw conclusions. Giving straw to calves that are fattened to such high final weights must help to increase welfare. There was some argument, at least in Germany, that providing straw would create another welfare problem, namely these ulcers, but we could not support this argument. How does one weight these two welfare problems?

G. van Putten It is difficult to balance these two issues, but there is still another component in this matter. In our calves the number of ulcers and lesions in the abomasum increased when additional barley straw was given but, on the other hand, the rumen showed less parakeratosis. The rumen was thus more normal and the abomasum more abnormal, so it is difficult to weight all these components.

N. Steenkamer It is important in such cases to look at a large number of calves and the report of Dr. van Putten is very valuable because his results and observations were from so many calves. We have also observed quite a number of abomasums and sometimes you find groups of 100 calves on a milk diet kept on slatted floors where there is not one calf with an abomasal ulcer. There are groups of calves kept under the same conditions where the percentage can be as high at 40%. There must be other important factors. The use of straw should not be criticised too much in regard to this problem.

There is an enormous variation in the percentage of abomasal ulcers found in different groups of calves and up until now no one has provided a good explanation for this.

W. Goldhorn *(CEC)* Dr. Smidt mentioned foot lesions and injuries on straw bedding. What sort of lesions were found and is there an explanation for their occurrence?

D.T. Smidt These investigations were by a veterinary pathologist working in Berlin. He found numerous haemorrhages in the papillary body of the feet of calves kept on straw and gives the explanation that the horn of the claws is weaker and more succulent and so the feet are more susceptible to such micro-lesions which lead to these haemorrhages. However, the calves do not seem to be lame or to find walking painful; these are histological findings.

W. Goldhorn That almost answers my second question. Do you then think that these lesions have any relevance to welfare since the calves do not suffer?

D.T. Smidt This is difficult to answer and one can ask the same question about abomasal ulcers, about parakeratosis in the rumen and the same about the lesions found in the joints of pigs and fattening bulls. However, if we agree that such lesions do cause pain although not severe pain then, yes, it is a welfare problem but we have little chance of really proving this because we cannot measure the amount of pain. However, there are no clinical symptoms of lameness with this.

D.B. Stephens *(UK)* I would like to ask Dr. Grongnet, or anyone else here, whether they have much experience of use of respiratory stimulants in calves which have had a difficult parturition?

J.F. Grongnet *(France)* No, I have never used such a thing for the improvement of respiration. I wanted to characterise anoxia and slight asphyxia. My work was not to improve practical methods for calf survival.

PHYSIOLOGICAL PARAMETERS WITH RELEVANCE TO CALF WELFARE

H.M. Eichinger

Aus dem Lehrstuhl für Tierzucht
der Technischen Universität München,
8050 Freising-Weihenstephan, Federal Republic of Germany.

ABSTRACT

Physiological parameters can be useful tools in the evaluation of calf welfare. However, it is necessary that the vitality and adaptability of the animals are known, that the environmental factors including circadian rhythms are well monitored, and that the criterion is sensitive to the influences being investigated.

Heart frequency is a recommended parameter in order to monitor directly the calf's reactions. For the estimation of physical adaptability a cardiorespiratory index is proposed.

As no accurate limits are applicable, investigations based on physiological parameters should have special experimental designs.

Welfare of an animal is difficult to define, more difficult to describe and even more difficult to measure.

It is generally agreed that environmental circumstances should allow an animal to keep all its physiological functions within a normal range. However, it is complicated to make a clear demarcation between the pathological and the normal range, and it is even more difficult to establish clear physiological differences between the less and the more comfortable animal.

A general view on physiological parameters in this respect is given elsewhere (Eichinger, 1976) and the following consideration concentrates upon the criteria of the cardiac and respiratory system.

The value of these physiological parameters in welfare evaluation depends to a large degree on:

- the animal, its vitality and its adaptability;
- the general environment, including training and circadian rhythm; and
- the parameter itself, which might be very sensitive to some factors and totally insensitive to others.

Physiological criteria as well as intrinsic and extrinsic factors can act in different directions involving more or less pronounced interactions and reflecting both more or less welfare (Figure 1).

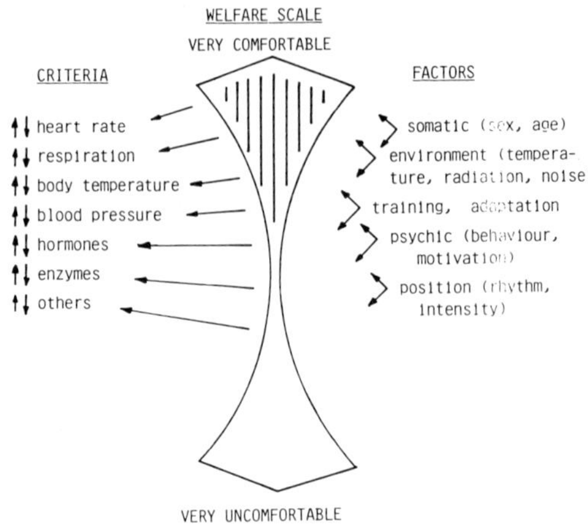

Fig. 1. Some physiological criteria and determining factors in calf welfare evaluation.

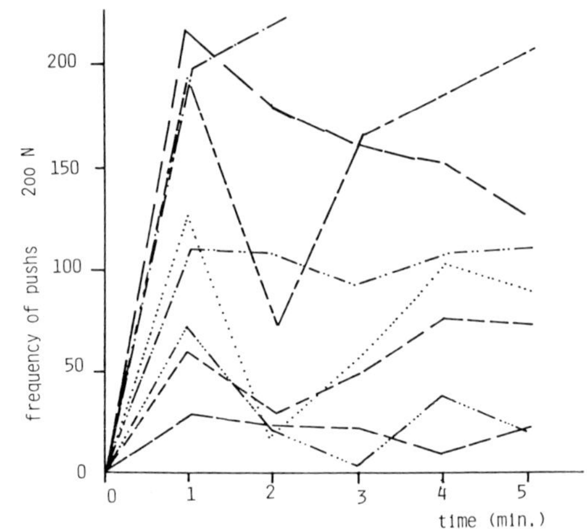

Fig. 2. Individual reaction of calves during first treadmill exercise.

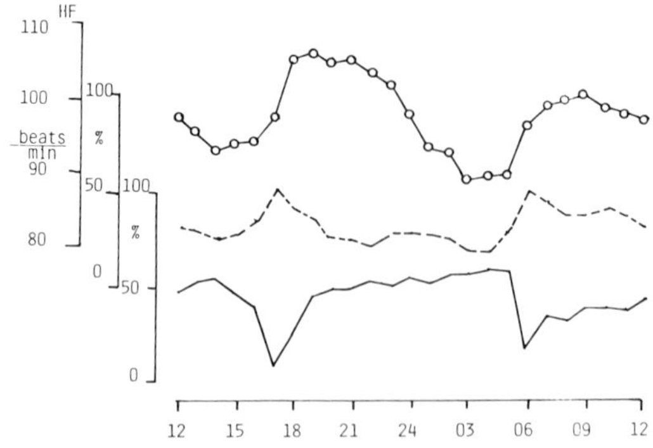

Fig. 3. 24 h pattern of heart frequency (HF), motor activity (A), and lying frequency (L) of individually caged calves (n = 44).

As regards the vitality and adaptability of an animal, very important physiological reference values are associated with physical activity. The latter can well be monitored by walking the animal on a treadmill, which involves some learning process and co-operative behaviour.

These individual abilities can be measured during first minor treadmill work by counting electronically the push frequency of the walking animal against the rear limitation of the treadmill (Figure 2).

The individual variation is remarkable, although this can be greatly reduced by assisting and leading the calf during the walking period. However, this involves a further source of variation due to the individual technician, so the number of uncompleted or insufficient treadmill tests can vary within the same group of calves from 0 to 15%.

If one or more environmental factors are subjected to change, the remaining extrinsic factors should be well recognised including the 'Zeitgeber' in circadian rhythms.

A circadian rhythm in nearly every physiological criterion is more or less pronounced in resting calves. An associated pattern can be observed when the telemetrically registered heart frequency, the electronically monitored physical activity and the opto-electronically measured lying frequency are compared (Figure 3).

Besides these similarities which can be seen at a first glance, the coefficients of correlation between heart rate and the activity-indicating parameters vary from hour to hour; they are highest at 0100 h (activity) or 1100 h (relative lying time), and lowest at 0500 h and 0700 h, respectively (Eichinger, 1980). These findings indicate strongly that even in daytime with much physical activity, most of the heart rate variation can depend on other than motoric factors, for instance digestion or emotional stress.

A recent study on transportation stress in calves (Niedermaier, 1981) showed no significant differences due to genotype (high level Brown Swiss crosses versus local Braunvieh), due to age (8 - 15 weeks) or to feeding intensity (±25%), but fasting longer than 14 h did reduce heart rate significantly during transportation. Further differences are due to several transportation stages producing very different reactions in the individual calf; the overall repeatability of the telemetrically monitored heart frequency from one transportation to the other was 40% to 60% (Figure 4).

Therefore, the adaptation performance of a calf is largely determined by the degree of physiological reaction. When one animal is suffering stress, the other may feel comfortable although under the same conditions.

For the estimation of physical performance, the regression coefficient of the heart rate on oxygen consumption per kg body-mass seems to be useful (Eichinger, 1980).

It is, in fact, very difficult to provide physiological standards for the proper evaluation of welfare or stress situations. Therefore, investigations on calf welfare by means of physiological parameters should be based on control groups and/or 'switch-over' experimental designs.

REFERENCES

Eichinger, H.M., 1976. Assessing husbandry responses on the basis of physiological parameters, Züchtungskunde 48, 437-446.

Eichinger, H.M., 1977. Untersuchungen zur Aktivitätsbeurteilung bei Belastungssituationen. EAAP, 28. Ann. Meeting, M 309.

Eichinger, H.M., 1980. Cardiorespiratory reactions of adaptation in calves. Inaug. Thesis Techn. Univers. Munich.

Niedermaier, H., 1981. Investigations on the influence of breed, type of nutrition, nutrition intensity and fastening on reactions of the circulatory system of calves during treadmill testing and transportation. Thesis Techn. Univers. Munich 1981.

THE LYING BEHAVIOUR OF VEAL CALVES UP TO 220 kg

G. van Putten and W.Y. Elshof
Research Institute for Animal Husbandry 'Schoonoord', Zeist, The Netherlands.

INTRODUCTION

Veal production in the EC amounts to about 800 000 t per annum (Animal Production, 1981) and is concentrated mainly in France, Italy, the Netherlands, West Germany, Belgium and in the United Kingdom. In the Netherlands (Mestkalverenbesluit, 1961), the United Kingdom (Codes of Recommendations, 1969) and West Germany (Kälbergutachten, 1973) a growing interest in the welfare of these calves can be noted.

In the context of this paper the expression 'well-being' is understood as defined by Lorz (1973). The evaluation of welfare is based on a philosophy as presented by Van Putten (1981). Thus we accept, that in every system of animal husbandry, and especially in intensive farming systems, producers make profit from keeping livestock on a commercial basis. We also accept the fact that (modern) farming systems generally have in some way or another a detrimental effect on the well-being of the animals involved.

The production of veal is not only economically important for the specialised veal calf farmers, but also for the majority of dairy farmers. In order to receive a fair price for a male calf from a dairy farm, the principal criterion is its fitness to serve as a producer of veal and by that we mean white veal. This whiteness has no influence whatsoever on the taste of the meat, it merely permits a classification of real white veal to be used. Incidentally, unless the meat is white, no market exists for this product in, for instance, Italy.

In principle only two types of housing are used for veal calves: individual housing and group housing. Individual housing can be realised in two ways: by keeping calves in boxes, or by tethering them between partitions. As a general rule, group housing is preceded by individual housing for a couple of weeks. Group housing is typical for Switzerland, southern Germany and eastern France. In the United Kingdom experimental farms for group housing exist. In other countries within the EC individual housing is practised. The problems mentioned in this paper originate from intensive systems with individual housing in boxes; however, I am sure some will also apply to group housing.

PROBLEMS

Like most livestock kept intensively, veal calves suffer from many problems. The severe anaemia (about 7 g % Hb as an average) at the time of slaughter certainly presents important economic factors. Anaemia at 20% (Van de Braak, 1978) in all newborn (0 - 3 weeks old) veal calves is entirely unnecessary and the problem should be treated in a proper way.

One of the greatest disadvantages of intensive farming systems centres round the lack of activity for the animals involved and this applies to veal calves as well. Their boxes, with slatted floors, are very barren; indeed, some box types do not even allow calves to look out as the front section is completely closed instead of being an open slatted construction. Sometimes veal calves cannot see out of their boxes at all because some fattening houses have no windows and artificial lighting is considered to be too expensive except during feeding times. Some countries have forbidden this situation, but in practice inspections are rarely carried out. Other bovines suffer less boredom than veal calves because feed intake and ruminating take up most of the time. Veal calves have no chance to eat roughage or to ruminate. It is clear that they miss this opportunity, because feint-ruminating can be observed for about 2 h/day . The most obvious way to provide some activity is by feeding some roughage which would enable the calves to ruminate - 75 g of barley straw/calf/day provides about 4 h of activity (van Putten and Elshof, 1978).

Being sucklings of social ruminants, veal calves are denied sucking and ruminating and they are also denied the company of other cattle. Every day veal calves try to reach one another and this they do at feeding times when the fronts of the boxes are opened to allow bucket feeding. They also try to touch, or at least see each other, by turning around in their narrow boxes (quite an effort for older calves) and to put their heads out of the rear ends of the boxes, where there are large gaps between the planks.

However, one of the most primary provisions veal calves lack is the opportunity to lie at their ease, and to get up and lie down in the specific manner of the species. This is one of the major animal demands indicated in the Brambell Report (1965) and as such recognised as one of the five freedoms of movement. This provision was also prescribed in the Dutch Veal Calf Regulations (Mestkalverenbesluit) of 1961. However, both the recommendations of the Brambell Committee and the Dutch Veal Calf Regulations in the Netherlands have been abused up till now. There are, of course, some reasons for this attitude. One of them is that the regulating paragraphs did not mention dimensions in centimetres for the whole range of calf weights or girth measurements.

AIMS

Confronted with the Dutch Regulations (Mestkalverenbesluit, 1961), the Brambell Report (1965), the British Codes of Recommendation (No. 1, Cattle) (1969) together with the German Experts' Report (Gutachten, 1973) regarding the welfare of veal calves, we decided to set up an experiment in order to correlate the views expressed in the documents mentioned above, using the exact dimensions recommended for the boxes.

The first experiment took place in 1974. On six occasions we bought a group of 27 representative male veal calves in the weight class 180 - 200 kg. They remained in the experimental fattening house for three weeks and were then slaughtered. Before being put in the experimental stall, where they were tethered at distances of 100 cm, they were kept in commercial fattening houses with boxes measuring 60 x 160 cm. The main conclusion (Van Putten and Elshof, 1975) from this experiment was that veal calves up to 200 kg need boxes with a width of at least 70 cm. However, there was one problem. Our calves might have been deterred from using the entire available space since they had been restricted to a much smaller area in the previous months.

Thus another experiment was planned in which 187 representative male newborn calves (about one week old) were again tethered at distances of 100 cm. The calves belonged mainly to the Dutch Friesian breed, although there were some Meuse, Rhine and Ijssel breeds as well, thus reflecting the practical farming situation in the Netherlands. Every two weeks they were weighed and measurements were taken in order to find the relation between bodyweight or girth and the area needed for lying comfortably and getting up without difficulty.

We wanted a whole range of relationships, beginning at 50 kg and ending at 220 kg, i.e. the slaughterweight. The girth and other body measurements were included in order to make inspection easier in practice. On a commercial farm it is technically almost impossible, and from the welfare point of view very cruel, to force calves out of their narrow boxes, weigh them and then force them backwards into the boxes again and onto the high slats.

OBSERVATIONS

If not indicated otherwise, all data originate from our experiment carried out in 1975 and published in 1980 (Elshof and Van Putten).

Lying time and lying posture

Graf et al. (1976) and Van Putten and Elshof (1975) agree that veal calves lie for about 15 h/24 h. During this time they lie for about 30 min completely on their sides, with the legs stretched out and the head on the floor (Van Putten and Elshof, 1975). These 30 min usually consist of 6 short periods each of 5 min.

The position of lying completely stretched out on the side may be important for the calf although we have no information about this. However, if we regard this position as being decisive, the minimum dimension for the width of a box is the croup height of the calves. Transferred to a calf of about 220 kg this would mean about 105 cm. Compared with a dairy cow of 500 - 600 kg, which has no more than about 120 cm at its disposal, it becomes clear that demanding 105 cm for a veal calf is not really practical. We decided to restrict ourselves to the other lying position that occurs for 14½ h of the total lying time of 15 h. In this position the calf lies with its breast between the (bent) forelegs and with the hindquarters stretched to the right or to the left side and forward. Only in this particular lying position are they able to belch and to ruminate. Within the framework of the Dutch law we considered that such a posture should be possible, as the width of the box must allow a calf sufficient space to do this. Scheuermann (1971) studied the lying postures of calves between 3 and 5 weeks old and distinguished no less than 40 different positions. Most of them were variations of lying on the breast, with a few lying on the side. Both Scheuermann (1971) and Graf et al. (1976) came to the conclusion that a calf can only lie as it would like to in a straw yard, which means group housing.

Lying down and getting up

Lying is one thing, but changing the position from lying to standing or vice versa is another. In order to change position cattle go onto their knees (or better, wrists) and lean forward so that as much weight as possible is shifted from their hind legs. They then get their back up (or down) in one swinging movement. In order to lean forward calves need about 25 cm and if this is added to their body length we can calculate the minimum length of the box required.

However, if this extra 25 cm is not available calves have to change position with the full weight on their hindquarters. In this situation swinging movements are out of the question and calves jack themselves up (or down) foot by foot. It will be clear that slippery floors add to the difficulties described.

Influence of environmental temperature

Even when lying on the breast the occupied area of the box varies considerably. This is partly dependent on the weight (or age, or size) of the animal, and partly on how far the hind legs are stretched outwards. The extension of the hind legs appears to be influenced by environmental temperature:

the warmer it is, the further the legs are stretched out. Because the summer of 1975 was a hot one in the Netherlands and because we had recorded temperatures in many areas we were able to study the relationship between room temperature and the area occupied.

Up to 180 kg bodyweight the regression co-efficient between environmental temperature and stretching of the hind legs was linear but different for each weight class (10 kg). Above this weight the function appeared to be square. In order to avoid extremes we did not allow temperatures in the occupied areas to rise above 22°C, this being the average daily maximum temperature in the Netherlands during July and August of that year.

The kind of data and the way these were obtained

Every two weeks, during the whole of the following period, a complete set of data was recorded for each animal. This meant weighing the animals on a set day and measuring body length, height of shoulder, height at croup, hip width and girth in the same week. The following week was taken up with measuring the distance between the maximum curvature of the back and the outer tip of the hoof, using special measuring rods painted matt black for the occasion. This distance was called the lying breadth. It was measured by two very skilled men and repeated five times for each calf. It was decided that these measurements should be repeated only after the calf had stood for some time. To make sure of this, at least one feeding was given between each measurement. These two experts had to be well acquainted with the behaviour of calves; they had to know how to move slowly and how to apply special rules in order to avoid the calf drawing in its hooves at the last critical moment whilst being measured. If this did happen then another attempt at measurement had to be made after the next feeding.

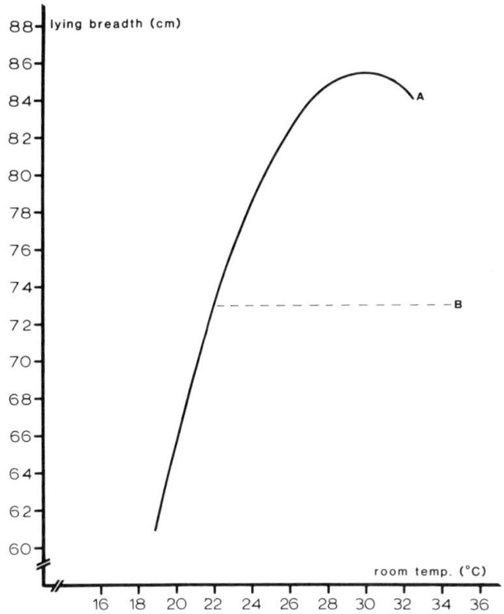

Fig. 1. Relation between room temperature and the occupied lying area for calves weighing between 210 - 220 kg (A). Line B has been drawn to illustrate the effect of reducing temperatures above 22°C to the average daily maximum temperature of 22°C during the months of July and August in the Netherlands. (Elshof and Van Putten, 1980).

RESULTS

The weights and girths are conveniently converted into classes. In the first two Tables the five measured lying breadths are arranged in size order, the smallest first, the mean in the centre, and the largest in the end column. The measured distances are not yet corrected for very high temperatures. Average values and their standard deviations are presented.

After correction for high temperatures, we restricted the data to the means of the lying breadth. Figures 2 and 3 show the results of this operation.

By restricting the lying breadths to mean values only and by reducing values measured under high temperatures, the data were considerably simplified. Table 3 presents bodyweight, mean lying breadth, girth and the calculation factors from bodyweight to girth in one convenient survey.

TABLE 1

THE RELATION BETWEEN WEIGHT AND LYING BREADTH (WITHOUT CORRECTION FOR TEMPERATURE). AVERAGE VALUES AND THEIR STANDARD DEVIATIONS OF THE MEASURED DISTANCES ARE PRESENTED IN cm, AND ARE ARRANGED AS PROGRESSIVE SERIES. THE BODY WEIGHTS ARE CONVERTED INTO CLASSES (ELSHOF AND VAN PUTTEN, 1980).

Weight	Breadth I (smallest)		Breadth II		Breadth III (mean)		Breadth IV		Breadth V (largest)	
(kg)*	av.	s.d.	av.	s.d.	av.	s.d.	av.	s.d.	av.	s.d.
70	52	3.5	56	4.2	59	4.5	64	5.6	70	6.0
80	53	3.3	57	3.9	61	4.3	65	5.1	72	5.5
90	55	3.4	59	3.8	62	4.4	67	5.2	73	5.0
100	56	3.6	60	4.0	64	4.8	68	5.5	75	5.3
110	58	3.6	62	4.2	65	4.9	70	5.5	76	5.5
120	59	3.5	63	4.1	67	4.7	71	5.4	77	5.4
130	61	3.5	65	4.1	69	4.6	73	5.3	79	5.4
140	62	3.4	66	4.0	70	4.4	75	5.1	81	5.1
150	63	3.4	68	3.8	71	4.4	76	5.0	82	5.0
160	65	3.5	69	3.9	73	4.4	77	4.8	83	4.8
170	66	3.6	70	3.8	74	4.3	78	4.7	84	4.5
180	67	3.8	71	3.8	75	4.1	79	4.5	84	4.3
190	68	3.9	72	3.9	76	4.1	80	4.8	85	4.5
200	69	4.2	74	4.2	77	4.6	80	5.6	85	5.4
210	70	4.4	75	4.6	78	5.8	81	6.7	86	7.1

* converted into classes.

TABLE 2

THE RELATION BETWEEN GIRTH AND LYING BREADTH (WITHOUT CORRECTION FOR TEMPERATURE). AVERAGE VALUES AND STANDARD DEVIATIONS OF THE MEASURED DISTANCES ARE PRESENTED IN cm, AND ARE ARRANGED AS PROGRESSIVE SERIES. THE GIRTHS ARE CONVERTED INTO CLASSES (ELSHOF AND VAN PUTTEN, 1980).

Girth	Breadth I (smallest)		Breadth II		Breadth III (mean)		Breadth IV		Breadth V (largest)	
(cm)*	av.	s.d.	av.	s.d.	av.	s.d.	av.	s.d.	av.	s.d.
85	49	4.9	54	5.6	57	6.1	62	7.7	69	9.9
90	51	4.0	55	4.7	58	5.1	63	6.4	69	7.9
95	52	3.6	56	4.3	59	4.7	64	5.6	70	6.4
100	53	3.4	57	4.0	61	4.4	65	5.1	72	5.4
105	55	3.5	59	3.9	63	4.4	67	5.1	74	5.1
110	58	3.7	62	4.0	65	4.8	70	5.5	76	5.4
115	60	3.8	64	4.2	68	4.8	72	5.3	79	5.3
120	64	3.7	68	4.2	72	4.8	76	5.2	82	5.0
125	67	3.7	71	3.9	75	4.1	79	4.8	84	4.5
130	69	3.9	73	4.1	77	4.5	80	5.7	85	5.4
135	70	3.7	75	4.5	79	5.6	80	7.3	85	9.0

* converted into classes.

TABLE 3

RELATION BETWEEN BODY WEIGHT, MEAN LYING BREADTH, GIRTH AND THE CALCULATION FACTOR FROM BODY WEIGHT TO GIRTH, AFTER CORRECTION FOR TEMPERATURES ABOVE 22°C (ELSHOF AND VAN PUTTEN, 1980).

Weight (kg)*	Mean lying breadth av. (cm)	s.d.	Girth av. (cm)	s.d.	Calculation factor weight girth
70	59	4.2	95	2.3	1.36
80	60	4.1	100	2.2	1.25
90	62	4.3	104	2.2	1.16
100	63	4.4	108	2.3	1.08
110	64	4.5	111	2.2	1.01
120	65	4.2	114	2.1	0.95
130	67	4.1	116	2.0	0.89
140	68	4.0	118	2.0	0.84
150	69	3.9	120	1.9	0.80
160	70	4.0	122	2.0	0.76
170	71	3.9	123	2.1	0.72
180	72	3.8	125	2.3	0.69
190	72	3.8	127	2.3	0.67
200	73	4.3	129	2.4	0.66
210	73	5.1	131	2.6	0.62

* converted into classes.

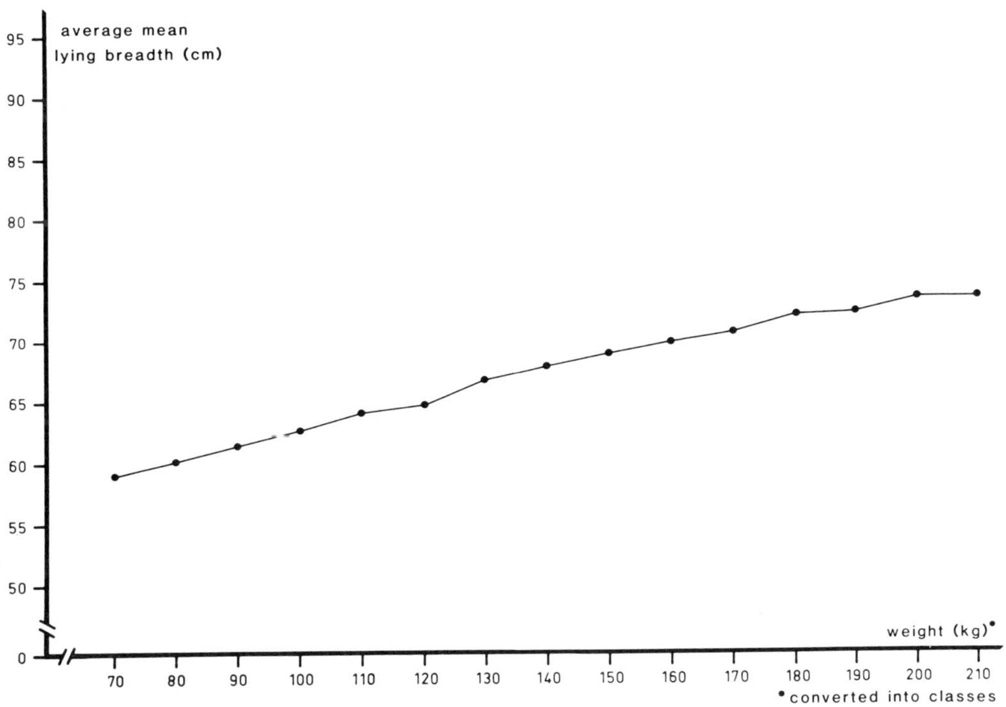

Fig. 2. Relation between lying breadth and body weight after correction for temperatures above 22°C (Elshof and Van Putten, 1980).

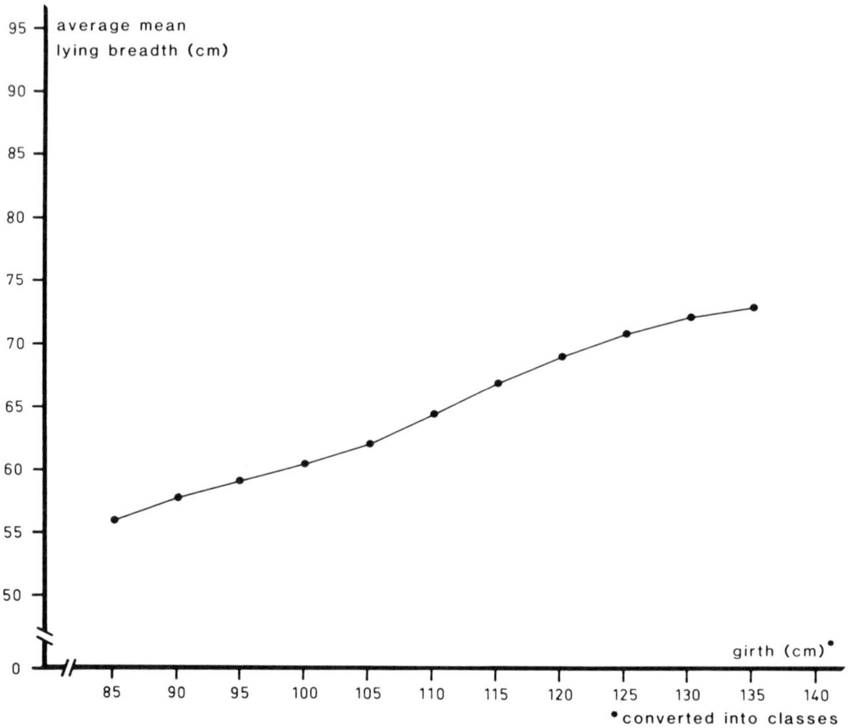

Fig. 3. Relation between breadth and the girth after correction for temperatures above 22°C (Elshof and Van Putten, 1980).

The relation between girth and bodyweight is expressed as a graph in Figure 4. Both are converted into classes. The availability of such a graph facilitates the estimation of bodyweight by a welfare inspector, who cannot actually weigh the calves.

In Figure 5 some other body measurements are presented. It will be clear that croup height, shoulder height and the hip width give only a poor indication of the animal's weight. The increase between weight classes is too small. Body length would have been of more help, if only this measurement could be carried out in a sufficiently accurate way. However, measuring a calf from tail to forehead is a very difficult job, and standard deviations of no less than 5 cm have to be taken into account. Nevertheless some indication of its value is important in order to design a calfbox with sufficient length to enable the animal to get up and lie down. Table 4 demonstrates the consequences of all these body measurements as regards box size.

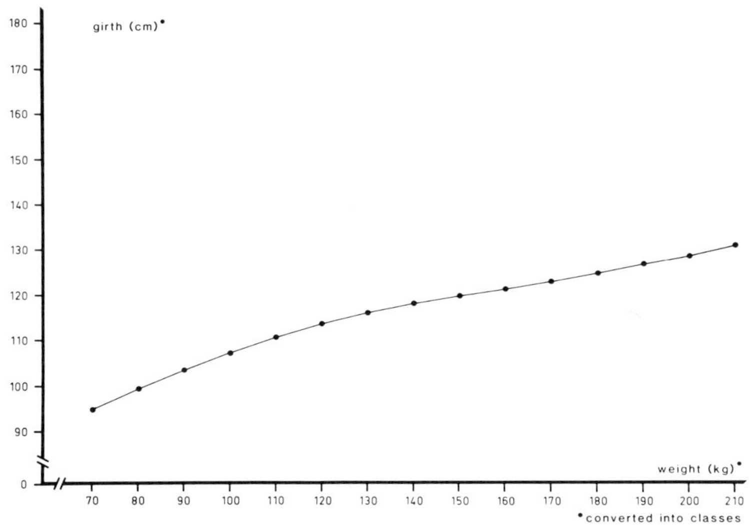

Fig. 4. The relation between girth and body weight. Both are converted into classes (Elshof and Van Putten, 1980).

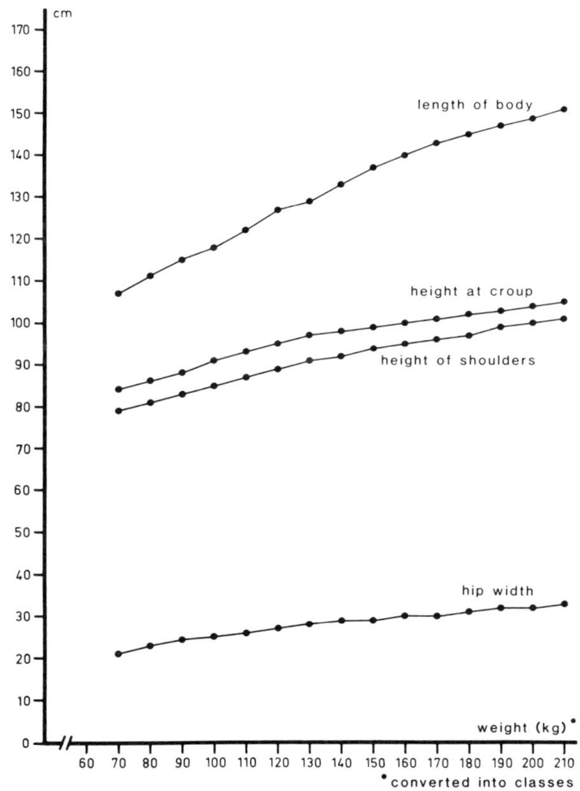

Fig. 5. Some body measurements in relation to the bodyweight. The body length is especially important in relation to box size. (Elshof and Van Putten, 1980).

TABLE 4

CRUCIAL BODY MEASUREMENTS OF THE CALVES AND THE CONSEQUENCES AS REGARDS BOX SIZE (ELSHOF AND VAN PUTTEN, 1980).

Body weight (kg)*	Girth av. (cm)	Body length av. (cm)	s.d.	Necessary length of box av. (cm)	Mean lying breadth av. (cm)
70	95	107	2.4	132	59
80	100	111	2.2	136	60
90	104	115	2.2	140	62
100	108	118	2.3	143	63
110	111	122	2.6	147	64
120	114	127	2.9	152	65
130	116	129	3.1	154	67
140	118	133	3.2	158	68
150	120	137	3.1	162	69
160	122	140	2.9	165	70
170	123	143	2.7	168	71
180	125	145	2.5	170	72
190	127	147	2.4	172	72
200	129	149	2.4	174	73
210	131	151	2.5	176	73

* converted into classes.

The results of Elshof and Van Putten (1980) are in full accordance with those of Van Putten and Elshof (1975), notwithstanding the fact that the 187 animals from the publication of 1980 were kept in unrestricted lying breadth during their complete lifespan, whilst those of 1975 (160 animals) were kept under these circumstances for only the last three weeks prior to slaughter. The 160 calves were kept in boxes (60 x 160 cm) before the experiment. Both publications indicate a box width of about 72 cm for veal calves with a weight of 180 - 200 kg.

In four successive trials Jongebreur and Smits (1978) compared the effect of the box widths (55, 60, 65 and 70 cm) on production and on some behavioural aspects. They kept about 70 calves in each category of box width. The differences in average growth over the whole fattening period of 20 weeks were small (about 1%). For the last month of the fattening period this increased to 4 - 5%. Only the growth of the calves in the smallest boxes was significantly different. Observations (slow motion projection) showed that lying down, getting up and grooming became difficult for calves at an age of 16 weeks (weight 170 - 180 kg) in boxes with a width of 55 cm. At an age of 18 weeks, the same situation developed for calves in boxes with a width of 60 cm. At an age of 20 weeks (weight 208 - 217 kg) Jongebreur and Smits (1978) came to the conclusion that only boxes with a width of 70 cm and a length of 170 cm ensured some freedom of movement for veal calves.

DISCUSSION

This paper is not entirely objective - choices have been made and priorities accepted. Although this can hardly be avoided in applied science, the subjective approach should be made clear.

- The possibility for a veal calf in a box to lie completely on its side has been dismissed, for economic reasons.
- Unimpeded lying (on the breast) is only possible in 60% of the situations (mean value) and then for only 50% of the calves (average lying breadth).
- If weather conditions make the room temperature in fattening houses in excess of 22°C, our calculations are no longer applicable.

On the other hand, difficulties with box size occur as a rule only in the last month of the lives of the veal calves involved, thus a box regarded as too small in these circumstances might provide sufficient room for calves in their first three months of life.

The question of whether unrestricted lying is a condition *sine qua non* for well-being, or that some restrictions do not really decrease well-being, remains open. In any case it was not our aim to prove either one point or the other.

CONCLUSIONS

In the Netherlands, West Germany and the United Kingdom a growing interest can be observed in the welfare of veal calves. Unrestricted lying is regarded to be an essential part of the calf's well-being. If interpreted literally the croup height should indicate the box width, because only then are calves able to lie with their hindlimbs stretched out completely. This would mean that calves with a body weight of more than 160 kg need a box width of more than 1 m. Because such a demand would not be reasonable for situations in practical farming, 'unrestricted lying' was described as lying with the breast between the knees and with the hind legs forward and to the side. The distance between the maximal curve of the back and the outer tip of the hoof of the hind legs was called 'lying breadth'. The lying breadth was regarded as decisive for the box width. Because the lying breadth appeared to be correlated positively with the room temperature, measurements of the lying breadth above the daily average maximum summer temperature in the Netherlands (22°C in the months July and August) were reduced to those at 22°C.

Every two weeks the weight, the shoulder height, the croup height, the body length, the hip width and the girth were measured. During every second week the lying breadth of each calf was measured five times. The mean value was the only one taken into account. The average mean lying breadths are presented in relation to classes of body weights and classes of girths. The latter is important for the purpose of practical inspection because the girth can easily be measured, whereas weighing is almost impossible.

The lying breadths of an experiment (Van Putten and Elshof, 1975) in which the animals were tethered at distances of 1 m having been kept in boxes with a width of 60 cm until a weight of 180 kg, showed no marked differences with those of a later experiment. In this last experiment (Elshof and Van Putten, 1980) veal calves were kept tethered at distances of 1 m from the age of one week until slaughter at 220 kg. In these two experiments 347 male veal calves of the breeds Dutch Friesian and Meuse, Rhine, IJssel were involved. The overall conclusion is that veal calves with a body weight of 200 kg need at least a box width of 70 cm and a box length of 170 cm. However, difficulties with smaller box sizes arise particularly in the last month of the fattening period. These difficulties concern restricted lying, getting up and lying down without great effort, as well as grooming.

REFERENCES

Animal Production, 1980, 1981. Luxembourg, Statistical Office of the European Communities. p. 86.

Van de Braak, A.E., 1977. Bloedwaarden samenhangend met ijzerarme voeding. Utrecht, Vakgroep Zoötechniek, Faculteit der Diergeneeskunde.

Brambell, F.W.R., 1965. Report of the Technical Committee to enquire into the welfare of animals kept under intensive livestock husbandry systems. London, Her Majesty's Stationery Office.

Codes of Recommendations for the Welfare of Livestock, 1971. Code No. 1: Cattle. London, Ministry of Agriculture, Fisheries and Food.

Elshof, W.J. and Putten, G. van, 1980. De ligbreedte van vleeskalveren in afhankelijkheid van gewicht, borstomvang en staltemperatuur. Zeist, IVO 'Schoonoord'. Rapport B-158.

Graf, B., Wegmann, R. and Rist, M., 1976. Das Verhalten von Mastkälbern bei verschiedenen Haltungsformen. Schweiz. Landwirtsch. Monatsh., 54, 333-355.

Jongebreur, A.A. and Smits, A.C., 1978. Boxafmetingen voor vleeskalveren. Wageningen, IMAG Publ. No. 110.

'Kälbergutachten', 1973. Gutachten über die tierschutzgerechte Haltung von Kälbern in Aufzucht und Mast. Bonn, Bundesministerium für Ernährung, Landwirtschaft und Forsten.

Lorz, A., 1973. Tierschutzgesetz. Kommentar von A. Lorz. Munich, Verlag C.H. Beck.

Mestkalverenbesluit, 1961. Koninklijk Besluit 080961, jcto artikel 1, van de Wet op de Dierenbescherming. 's-Gravenhave, Staatsdrukkerij.

Van Putten, G. and Elshof, W.J., 1975. Platzanspruch eines Mastkalbes. In: Haltungssysteme und Verhaltensanpassung. Bericht über die Tagung der DVG, Fachgruppe Verhaltensforschung, Freiburg/Breisgau, October 1974. Darmstadt, KTBL. p. 56-63.

Van Putten, G. and Elshof, W.J., 1978. Zusatzfütterung von Stroh an Mastkälbern. In: Aktuelle Arbeiten zur artgemässen Tierhaltung 1977. Bericht über die Tagung der DVG, Fachgruppe Verhaltensforschung, Freiburg/Breisgau, November 1977. Darmstadt, KTBL. p. 210-219.

Van Putten, G., 1981. Parameters for farm animal well-being, emphasizing the role of conflict - and vacuum behaviour. 32nd Annual Meeting of the EAAP, Commission on Animal Management and Health, Zagreb.

Scheuermann, E., 1971. Die Ruhelagen des Kalbes. In: Verhaltensforschung beim Rind. Bericht über die Tagung der DVG, Fachgruppe Verhaltensforschung, Freiburg/Breisgau, November 1970. Frankfurt/M., KTBL. p. 100-118.

DISCUSSION

F.R. Bell *(UK)* These very interesting results remind me of some work done at The National Institute of Research in Dairying about twenty years ago. The study was in relation to sleep in cattle and it showed that because of the need to eructate it is improbable that adult cows ever wholly sleep. This was taken up subsequently by Jouvet in Lyons, France, and he went on to study paradoxical sleep. This is a parasympathetic condition where you have got radiation to the dorsal motor nucleus of the medulla and inhibition of heart-rate. This is spread over many species including man and the symptoms of paradoxical sleep are parasympathetic. In man one of the most common is penile erection, another is the slowing of heart-rate when you get palpitations at night. You also get movement of the face and rapid eye movements. Ruckebusch in Toulouse, France, showed that in newly-born calves there was a great deal of paradoxical sleep, rapid eye movement sleep, and as the calves grew older this diminished until they became adult and there was an abolition of this. Cattle thereafter adopted this sternal recumbency that you mentioned so that they could regurgitate. Was lying on the side early on in the veal calves in fact paradoxical sleep? Did you notice any movement of the eyes, face or limbs?

G. van Putten *(Netherlands)* No, we did not, but this does not mean that it has not happened because we were never very close to these animals. We had them tethered and walked by them to make quick notes of what was happening. It might well have been that some calves showed these symptoms but we did not observe them.

A.J. Webster *(UK)* I think it is appropriate to mention this now although I will return to it in my own paper. I have worked with Ruckebusch's team and measured paradoxical sleep with EEGs and the like, but it is in fact quite easy to observe from rapid eye movement measurements. In support of what Dr. van Putten has said, we would claim that the main constriction on behaviour in a box that is too narrow is not that it denies the animal the right to lie on its side. This only becomes a problem in practical terms when the environment is very hot since the animal does this for heat dissipation reasons. When a veal calf, in a conventional box of 65 cm or less width, reaches about 14 weeks of age it cannot adopt the classical sleeping posture with its head turned round on its side - not because the calf has got wider but because the neck has got longer. We have evidence in these older calves of a depression in total sleeping time and also a depression in clearly recognisable rapid eye movement sleeping time. In my opinion the greatest deprivation to the calf in the too narrow box is the deprivation of normal sleep.

GENERAL DISCUSSION

R. Moss *(UK)* There were two papers dealing with behaviour, two with physiological parameters and the recording of those, and three on both behaviour and physiology. Else Hermann and Dr. Le Neindre started off with the behaviour of the mother and calf. Dr. Stephens with his fascinating talk introduced neurochemistry, and in particular the renal blood flow. Dr. Grongnet spoke of very interesting measurements of various biochemical parameters in the blood. Dr. Smidt's was a mixed paper in relation to observations on behaviour and biochemical parameters. Dr. Eichinger also warned of the assessment of the parameters that were recorded. Finally, Dr. van Putten's paper which dealt with both behaviour and measurement, perhaps not the usual measurement but measurement of the physical calf itself.

I have been struck by the way in which normal selection procedures, the pressures of the industry, have changed the genetic type so that the animal's reaction to its environment has changed markedly from its very near genetic ancestors. It is now more adaptable, as was pointed out to us in one of the papers. Renal blood flow which may or may not have an apparent correlation with the welfare of the calf is an interesting parameter and one wonders if there is anyone searching for other parameters of that sort. We have been looking for many years at heart-rate, blood physiology and blood biochemistry.

H. Eichinger *(FRG)* There are a lot of other parameters. I think it is not as useful to measure renal blood flow when you now have very different sorts of animals. For instance, the normal physical capacity or performance of calves from the same group or herd may vary as much as 30 - 40%. If you measure just one criterion then you see only one small side of the problem. I think welfare is very complex and if you just consider one criterion then you must make a very good analysis of variance. I do agree that we could have many more special physiological criteria; the work done by Ruckebusch is valuable but we must be careful with these parameters.

G. van Putten *(Netherlands)* Do you think that we need all these parameters? For instance, a calf in a box that is too short has to get up and lie down with full weight on the hind legs, thereby making the normal swinging movement impossible. Do we need such observations confirmed by renal blood flow, or heart-rate, or blood pressure or whatever? I think that some things are so obvious that it is not necessary to look for other parameters. From the physiological viewpoint, these parameters are very difficult to interpret. If we are measuring the way an animal reacts to stress then we do not even know whether we have to mark the results of our measurements with a plus or a minus. Stress, to a certain level, can be positive for animals, and judging whether an animal is overstressed can be very difficult. Stress levels can differ even within the individual animal depending on the situation and the environment.

R. Moss I think, Dr. van Putten, that you have answered the question for me. I would certainly agree with you that the calf should be allowed to get up and down normally in accordance with its own structure. However, you have moved over into other areas; there may be bad welfare simply in keeping a calf isolated. So, although there are very clear-cut parameters such as you speak about, there are others that are not so clear-cut.

F.R. Bell *(UK)* With regard to the measurement of blood flow in the kidney, I agree with Dr. Eichinger that we should be very careful in animal welfare experimentation. We must also be prepared not to run away from the word 'experimentation', we should be prepared to experiment. I think we must be careful to use non-invasive methods whenever possible. I cannot at the moment think of any way of measuring blood flow not using a non-invasive method. The more instruments that we can keep away from the animal and the more we observe and tabulate and bring together and use covariance and multi-factorial inter-relationships, then the more we will begin to make progress. However, I am very doubtful whether measurement of blood flow in the kidney is going to help very much.

W. Goldhorn *(CEC)* The measurement of renal blood flow would be very important if it could help us answer the question of whether normal behaviour is a matter of adaptation or a sign of suffering. This distinction is not at all clear.

D.B. Stephens *(UK)* The idea of measuring blood flow to the kidney started from some observations I made with Bernard Williams of the Ministry of Agriculture Veterinary Investigation Centre (VIC) in Carmarthen, South Wales. I also knew the proprietor of the abbatoir in Nantwich where these small 'Bobby' calves were taken. Quite often a few calves were dead on arrival at Nantwich and from the Ministry's investigations at Nantwich it was found that there were pathological lesions in the kidneys of these young calves. We looked then at the blood urea levels of a batch of calves at Carmarthen, at the point when they were put onto the lorry. They were sampled on arrival at Nantwich and we found that there was an increase in the blood urea levels. This may not be entirely due to a change in kidney function. There may be redistribution of body fluids with haemo-concentration which might account for these increased levels. However, this did lead us to think that the kidney was an important organ and that there may be a redistribution of blood from organs such as the kidney which may then be redistributed to the bulk of the muscles of the animal in order for it to be able to stand and cope with the journey.

I then looked up the literature on kidney function and got to know Professor Henry at the University of Los Angeles, California. He sent one of his collaborators to work with me at Babraham who brought with him the facility for measuring renal blood flow. He made the blood flow probes which I could surgically place in the calf or pig. That is why we did the work. It seems to be a very sensitive index of sympathetic discharge or activation. We can say that at this stage and that if you do in any way arouse the animal, change its emotional status, you will end up with a very subtle, small but detectable change in the rate at which blood flows to the kidney.

P.V. Tarrant *(Ireland)* What about the influence of dopamine on frustrated sucking behaviour in the calf? Is dopamine involved in any other behavioural activities in the calf? Are there activities in the adult animals under the control of dopamine release?

D.B. Stephens The answer as far as domestic animals are concerned is no. Much work on dopamine, which is a naturally occurring central transmitter, has been done with laboratory animals. I do not know of much other work. Some work has been done on the pig at Babraham. It has been shown that different regions of the pig's brain produce more dopamine than other regions. Half the pigs were allowed to suckle from a teat, the other half were fed intragastrically with tubes and so were frustrated from suckling milk. One would look at different regions of the brains of these piglets where there would be increased metabolites of dopamine like HVA. One did find a distinct change, an increase amongst those piglets that did not get their milk from the teat, i.e. those that were fed intragastrically. As far as calves are concerned, we have not really studied this area at all.

F.R. Bell Also in answer to this question, we have done a little work with dopaminergic myoteric plexuses in the calf and we have shown certainly that there are a lot of dopaminergic plexuses in the calf that control intestinal movement. Dopamine certainly produces alimentary 'hurry', it can produce diarrhoea in the calf. Of course, as Dr. Stephens has said, you get plenty of dopamine in the brain as well as in the alimentary tract.

H. Eichinger I would like to comment on the idea that modern animal production has changed the adaptation of the animal. The animal has adapted to modern circumstances but its capacity for adaptation has been reduced at the same time. It cannot adapt like the original animal, for instance, to more or less heat, or to more or less food deprivation.

P. Le Neindre *(France)* I do not think that anybody could define what an original breed is. When I take a Friesian it can be treated either as a Salers (mixed beef-dairy) or as a Charolais (beef). When I take a Salers it can be treated only as a beef breed and it cannot be raised under dairy conditions. These are the only two conditions and there are no more original breeds, these do not exist any longer. Some cattle are more adaptable than others however.

W. Sybesma *(Netherlands)* I would like to take this subject further. Would it not be a good idea to try to find tests which could elucidate the predisposition for various threats to welfare? There are apparently some breeds more susceptible to one thing and other breeds more susceptible to other things. What about Else Hermann's findings with calf and dam inter-relationships? Is it possible to evaluate which calves and dams had better welfare in comparison with the others?

Else Hermann *(Denmark)* The beef dairy cows and calves had the best welfare, but all the dams licked their calves and ate placenta. This was a kind of fixed action pattern and I believe that this type of behaviour cannot be adapted.

W. Sybesma Did you find differences in the genetic responses? If so, would it not be a good idea to find some test to assess the genetic predisposition?

R. Moss It is important to note how one interprets that record. In some ways it was unfair to ask Else Hermann about her classification of welfare within the dam and calf of the three particular groups with which she worked. It was a very subjective judgement that she had to give. There was really only one thing that she could say, that those animals that to her and probably to most of us here, were allowed to behave more 'naturally' than the others had better welfare. This might not have been the case but it is an obvious first judgement.

J.P. Signoret *(France)* I would like to mention some experiments made on sheep rather than cattle by Pascal Poindron in France, concerning maternal reaction at birth. He has observed that there is a critical period around birth under hormonal control when the female is strongly attracted, as Miss Hermann has reported in cattle, by the foetal fluids. This attraction exists for several hours before parturition. When the ewe is presented to a new-born lamb she will immediately react towards the placenta as if she had given birth to that particular lamb, despite the fact that she will soon be giving birth herself. She reacts to this lamb and its placenta exactly as after parturition, and will do the same after birth. This period is of very short duration and if the new-born is removed just at birth and if the female has no chance of being in contact with a wet new-born or a placenta during this short critical period, then all her maternal reactions begin to disappear. If the ewe has no contact with the placenta, the fluids and the new-born for say six hours then the female will no longer accept any new-born and will have to be forced to accept one. So, if we eliminate this period then no maternal drive will persist; the drive will only exist if there is a dialogue between ewe and lamb.

Concerning dam welfare, we can question whether there is a problem of welfare in leaving the calf for say 72 h with the dam. At the end of this period we disrupt the maternal bond completely. If there is no contact, then there will be no maternal bond and a completely different relationship exists. This could be an important issue for welfare, not for production. We could examine whether oxytocin is released and also examine the control of the mother's brain to see whether or not she has experienced this type of bond.

P.J. O'Connor *(Ireland)* With reference to the comment made by Dr. Signoret, shepherds have always used placenta to foster a lamb to a ewe. By draping the placenta over the lamb, the shepherd can encourage the ewe to accept the lamb. This is an ancient tradition and one that, as far as I know, is still being practised.

Having listened to the excellent papers today, I would like to say that if we come to translate welfare into legislation at some date, then should we be facing the question of the elimination of the veal box? Should the calf be put back into some social contact with its comrades? I think that we will have to approach this question decisively at some stage, despite the large investments and different approaches applied to veal production. Should animals have some possibility for social contact?

J.P. Signoret Concerning the necessity of social contact for the calves, I would like to mention a recent experiment by Marie-France Bouissou and performed at our laboratory. She kept calves totally isolated from other cattle. Eight calves were kept in barns without any possibility of receiving contact or stimuli from cattle. She observed the social interactions when these calves were placed together in a social group at the age of 6 - 7 months. In this way she could test the normality or the abnormality of their social behaviour. Abnormality was the criterion used to show how the animal has suffered in the social behaviour.

The most surprising thing was that isolated Friesian heifers were almost immediately regulating their social relationships. The major differences, which lasted only a few hours, were that they were less aggressive and exhibited more fear towards other heifers than normal socially-reared calves. However, within a matter of hours or days, they established mutual dominance relationships, a social organisation and normal frequency patterns of threats, moving away, food competition and so on. They became indistinguishable from normally reared animals within a few days.

From this, it can be concluded that social interactions return to normal very rapidly even in cattle which have been completely and thoroughly isolated. This has only been tried with Friesians, but Le Neindre has shown that there are some long-lasting consequences that appear among Salers but not among Friesians, so perhaps breed type accounts for differences. However, these results are very different from the complete abnormality that characterises the rhesus monkey kept without social contact.

N. Steenkamer *(Netherlands)* I would like to go back to the statement made by Dr. O'Connor on the possible effects of what has been said today on legislation. I think it is too early to make such decisions. When talking about animal welfare we have to deal with a set of priorities, but up to now we have not decided upon these priorities and in regard to this we must remember to take into account breed differences as well. We must also consider that when we criticise or condemn a certain system there must be a good alternative to replace it with. At the moment, the alternatives for the single pen (not the meshes) that give good results are not really available. I am not just talking about economics, although in our life economics must play an important role, but about welfare as well. To discriminate at this moment that no box is good for calves and that other systems providing more social contact between animals are by definition better would not be a good idea.

J.Y. Blom *(Denmark)* This is just a small comment on the social lives of calves. We noticed in our trials that boxes that were closed up to a height of 100 - 120 cm made the calves nervous. When the people attending the calves went into the stable to feed them, the calves were nervous, restless and often tried to jump into the neighbouring box. Such a reaction never appeared in calf houses where the boxes were closed up to 80 cm. Although no conclusions have been made on this, it seems that calves do become nervous when they are in very close confinement.

J.F. Grongnet *(France)* I would just like to add something about different housing facilities. Tomorrow morning, I will present the work of Jean-Paul Quillet who has not been able to come to this Seminar. His paper compares two types of housing: very light housing in which animals are kept in groups on straw, and classic, expensive housing that has cooling systems etc. No observation has been made about welfare specifically, but welfare is judged as good when daily weight gain is good, when no feed is refused, etc. From this you will see that the performance of the calves in the different housing systems is exactly the same. This was most interesting for Dr. Quillet because he works in practice and in France most veal calves are kept in closed, dark housing which is bad for the welfare of both animals and farmers. Farmers are anthropomorphic beings who feel sorry for their calves. We have found that in open, light housing the performacne of calves was very good.

R. Moss I will not ask anyone to follow up this point, we will wait till tomorrow, because I am sure that there are many here who would take issue with you for using production as almost the sole parameter of animal welfare.

H. Eichinger We have talked about the capacity for adaptability in animals and I think I would agree that the modern animal is not, at first glance, an animal with a greatly reduced adaptability. For instance, we compared normal local Braunvieh cattle with Brown Swiss imports and they proved to be very strong especially concerning the cardio-respiratory system. However, pigs behave quite differently from cattle.

There was a question from Dr. Sybesma concerning the measurement of performance, and that is what we try to do. We subject calves to a normal treadmill run for three periods of about four minutes, firstly in a horizontal position and then an inclined position. We then measure the heart-rate and oxygen consumption. We then make a regression from the heart-rate to the oxygen consumption and this gives us a certain idea of how strong the reaction was concerning the physical performance. I would not say that this will greatly help in determining how well the animal feels, but it does determine better performance. It would seem to be useful to check whether an animal is a good performer or not.

W. Sybesma Would Dr. Eichinger suggest different kinds of transport for animals that perform in different ways?

Do different groups also have different requirements for transport to the slaughterhouse? Should you adapt transport to suit the weakest?

H. Eichinger I would not draw any conclusions at the moment. There are animals that are very resilient that do not suffer under certain circumstances although other animals are under a great deal of stress in the same circumstances. You would not have to look at large groups of animals, you would only need to look at the poor performers and then look to see whether transportation or anything else can be improved.

G. van Putten There was a short discussion on the experiments of Mlle. Bouissou and her Casperhauser calves. Dr. Steenkamer stated that calves penned individually do not suffer. I think both of them are right, but the situations are really entirely different. Mlle. Bouissou kept the calves totally isolated from each other. When calves are individually penned one next to the other, the first thing that these calves try to do is to reach another calf. If this is impossible because the front is closed, made of slats or constructed in such a way that the calves cannot reach each other at the front, then they tend to turn around and try to contact other calves at the rear end of the box. I think this shows that the calves would like to be in contact and would certainly like to orientate, to have a clear view of their environment. Mlle. Bouissou did not allow her calves to see each other which is a very different situation, but when calves do see each other they try to come into contact.

R. Moss We have examined behaviour in a number of circumstances and in respect of certain biochemical, physiological parameters. We have not immediately drawn any conclusions from this behaviour or these parameters. However, it also seems clear that an animal is provided with a physical structure which allows it to do certain things in a certain way. It could be argued that it should be able to exercise those functions. It is also provided with neural templates which, as Dr. Signoret has pointed out in the case of the lamb and the ewe, need to be activated at certain times. This applies to the calf and the cow as well. If they are not activated at the right time they are never activated.

One of the key problems, if we accept Dr. van Putten's point about providing sufficient space to perform the physical acts that are required to be done, is that if you do not allow a neural template to be activated are you causing bad welfare? It is here that Dr. Stephen's interesting observation on the sympathetic system and the renal blood flow certainly is intriguing.

SESSION II

ENVIRONMENTAL PARAMETERS, HYGIENE AND CONDITIONS FOR WELFARE OF VEAL CALVES

Chairmen: H.C. Adler and W. Sybesma

SEVERAL MINIMUM REQUIREMENTS, FROM THE STANDPOINT OF ANIMAL PROTECTION, FOR HOUSING AND FATTENING OF CALVES

H. Bogner

Bayer. Landesanstalt für Tierzucht,
D-8011 Grub/p. Poing, Federal Republic of Germany.

ABSTRACT

As an introduction to the subject, the situation which shows a fall in numbers of fattening calves is described. Of the numerous ethological problems, four outstanding points are discussed:

1. *Measurements for stand length and width should be calculated on the basis of objective characteristics, i.e. the height at withers.*
2. *The floor surface of standing and resting areas should be non-slip.*
3. *It is not acceptable for calves to be fed exclusively fluids. From an ethological point of view, and to reduce the formation of hairballs, fattening calves should be fed at least a small amount of roughage. The causes of ulceration in the true stomach are not yet known (Bogner et al., 1981).*
4. *The sodium (Na) and potassium (K) content of the milk replacer is of great importance in avoiding hypernitraemia. On hot summer days an additional amount of water should be offered to fattening calves, or the concentration of the fluid feed reduced.*

This list of proposals is in the welfare interests of fattening calves and should be taken into consideration when laws for animal protection are drawn up.

INTRODUCTION

Every year in the Federal Republic of Germany about 650 000 calves, most of them female, are fattened with milk replacer and slaughtered at a weight of approximately 175 kg. Specialised calf fattening farms try to achieve a final weight of between 150 and 220 kg. Calves usually attain this weight at the age of 110 - 140 days.

It is believed that 90% of the fattened calves are produced in farms with less than 200 standings. Specialised farms buy their calves in bigger lots at auctions or from cattle dealers, whereas smaller fattening farms get their calves directly from the dairy farmers. The calves are normally bought at the age of 2 - 6 weeks and at a weight of 50 - 60 kg.

Preferred practice is to feed the calves from buckets and to keep them in single pens. In most cases the resting area is unlittered and perforated (having a slatted floor). Group penning on litter, favourable from an ethological point of view, is used only in exceptional cases with fattening calves. Group penning of calves is sometimes carried out on slatted floors. In group pens calves are usually fed in closed standings from buckets; automatic feeders are rarely used and only in farms with little livestock.

As a result of the recent 'oestrogen scandal', calf fattening has decreased further and veal prices are falling.

SPACE REQUIREMENTS

If calves are kept in pens or tied stalls, it is most important with respect to the welfare of the animal that sufficient floor space is allowed. The body measurements of the animal in resting, standing and feeding positions must therefore be the primary basis for rules and regulations relating to space requirements. Numerous experiments performed in single or group penning systems have shown that the width and length of the area available must be such that the animal is able to lie down sideways with its limbs stretched out. This resting position, although it amounts to only 10 - 15% of total resting time, is nevertheless a decisive factor in the choice of minimum dimensions of calf pens.

Space requirements are largely determined by the height at withers and by the total length of the animal which corresponds approximately to 1½ times the body length. Experiments carried out on 2 098 calves at the Bavarian Institute for Animal Breeding in Grub showed a high correlation ($r = 0.89$) between the height at withers and the body length.

It was therefore considered appropriate to introduce the height at withers, which is a measure with high repeatability, into an equation for estimating the standing area (y). This enables the rectangular form of the calf to be estimated by means of a relatively easily measurable characteristic and so provides a 'tailored' area for calf pens. On the basis of empiric observations, it has been found that the minimum length of the standing can be calculated by using 1½ times the height at withers (x) (Bogner and Alps, 1980, unpublished).

With respect to the resting position mentioned above (sideways with limbs stretched out), the width of closed pens which will assure a minimum of mobility is a function of the height at withers. In the case of partitions which leave an open space of approximately 25 cm above the floor so that calves can stretch their limbs, the width is reduced by a factor of 0.7.

The regression equation (Table 1) with the independent variable (height at withers) and the dependent variable (minimum area) can be used for final weights of up to 200 kg, as are common at present, regardless of the age or breed of the animal.

As well as formulae which estimate the size of closed pens and the size of pens with open partitions, a formula for the estimation of the area of group pens has also been developed.

STANDING AND RESTING AREAS

Calves kept in littered group pens have been found to stand more frequently than calves in tied stalls. Without a doubt, this is due to increased activity and higher mobility on non-slip flooring.

The standing period is shorter in littered group pens than it is in tied stalls. This leads to the conclusion that the urge or compulsion to move about, possibly provoked by social partners, can be more easily satisfied in group pens than in tied stalls.

However, an experiment in tied stalls with variable partitions (Reissig, 1979) showed the good adaptability of young calves. This proved that during the relatively short fattening period, calves are able to accept a certain space limitation without any decisive diminution of fattening performance. Nevertheless, it must be ensured that calves in tied stalls are able, at any time, to assume a relaxed resting position and to lie down and stand up freely.

It is regrettable that in recent years, for economic reasons, the solid dung method has been increasingly replaced by slurry disposal systems without litter. In numerous farms calves are kept on a slatted floor from the age of two weeks. From the ethological point of view, however, it is not desirable

TABLE 1

SPACE REQUIREMENTS OF CALF BASED ON HEIGHT AT WITHERS (x)

Height at withers (cm)	Liveweight of calves kg (approx)	Solitary rearing				Group rearing
		Without side walls		With side walls		Surface/calf (cm^2 and m^2)
		Length (cm) (0.5x + 45).1.5	Width (cm) x.0.7	Length (cm) (0.5x + 45).1.5	Width (cm) = x	138x + 640
80	67	128	56	128	80	11 680 = 1.17 m^2
90	123	135	63	135	90	13 060 = 1.31 m^2
100	180	143	70	143	100	14 400 = 1.44 m^2

TABLE 2

WATER REQUIREMENTS OF CALVES OFFERED FLUID FEED

		Beverage (l)	Milk replacer (kg)	Additional water *ad libitum* 20/2/79-1/5/79 4 - 13 week	Final weight (kg)	Daily weight gain (g)	Food Conversion
Group 1 n = 15	$\bar{x}$	1026.2	187.2	-	178.3	888	1.647
Group 2 n = 14	$\bar{x}$	907.3	194.8	-	177.1	902	1.743
Group 3 n = 13	$\bar{x}$	909.3	195.2	302.9	183.2	936	1.645

Group 1: milk replacer + straw - concentration 80 - 200 g/l (control group) without water (in addition)
Group 2: milk replacer + straw - concentration 80 - 270 g/l without water (in addition)
Group 3: milk replacer + straw - concentration 80 - 270 g/l + water *ad libitum*, 4 - 13 week of fattening period
Additional water was available from day 63 to day 70.

to keep fattening calves without litter before the age of six months. The use of litter could prevent many injuries caused by the form of housing (technopathy) and furthermore would diminish the energy input necessary for airconditioning in the stable.

Until this can be achieved, great attention must be paid to the structure of the floor surface in calf pens. It must allow calves to stand with safety and must be non-slip. Wooden floors, despite the good insulation they give, easily become slippery when wet with urine or dung and if no litter is laid, animals can easily slip and injure themselves. In contrast, rubber mats, plaster, grid or slatted concrete floors are non-slip surfaces. Recently, there has been an increasing use of perforated and, in particular, slatted concrete floors in the construction of stables. These floors can be recommended from the point of view of hygiene because urine and dung are removed quickly. Great attention has to be paid, from the point of view of animal welfare, to see that the concrete beams form an even surface, that the spaces between the beams are equal, that the width of the beams is not smaller than the animal's hoof and that the hoof cannot get caught between the beams.

In future, the aim should be to keep all fattening and rearing calves principally on littered resting areas. Other areas can without risk have floors of a solid material.

NUTRITION OF FATTENING CALVES

Once the suckling period is over, eating roughage and ruminating is a specific physiological function in calves. According to Porzig (1969) calves start to eat roughage and to ruminate at the age of 1 - 2 weeks and at 2 months of age they already spend 6 h/day ruminating.

Nowadays most fattening calves are exclusively fluid fed. This is intended to prolong the suckling period artificially and to suppress the development of the calves into ruminants with fully developed rumens. In the past, when calves were slaughtered at the age of 4 - 8 weeks, with a weight of about 80 kg, this method of feeding could be tolerated, but today's fattening calves reach 2 - 3 times this weight.

Exclusive fluid feeding results in calves showing 2 - 5 times more suckling activity on both living and dead objects. The reason for this is that the rumen has been insufficiently filled with bulky material (Groth, 1978). Furthermore, an increased number of Zootrichobezoare (hairballs) can be found in the stomach of fattening calves. The number of hairballs can be reduced considerably by feeding only 100 g chopped or soft straw, but the formation of ulcers may increase (Bogner et al., 1981). The causes of ulceration in the true stomach, which do not normally show clinically, are not yet known and several institutes are working on this problem.

Whether or not calves are fed straw makes no difference to the quality of the veal (Bogner et al., 1973). Daily weight gain is higher when straw is fed in addition to fluid. These findings were confirmed by Van Putten (1977).

WATER SUPPLY

Another welfare problem is that calves' water supply is limited by the quantity of milk feed. As the kidney has only a limited capacity for concentrating sodium (Na) and potassium (K), the secretion of these elements is dependent on the water supply being sufficient. As a result of the limited water supply, fattening calves may sometimes show hypernitraemia. Legislators have therefore decided that milk replacers must contain at least 2 000 mg K/kg dry matter and no more than 6 000 mg Na/kg dry matter. This is a sensible prophylactic measure (Gropp et al., 1978). Nevertheless, the wellbeing of the animal can be considerably affected if the milk replacer is not mixed correctly.

Table 2 shows the actual water requirement of calves when fluids with high concentration are offered. The calves in group 3, which were fed the same ration as those in group 2, had an additional water intake of about 300 l during the fattening period (4.3 l/animal/day). The calves in group 3 showed a higher daily weight gain and better food conversion than those in group 2. Therefore, at least on hot summer days, a reduction in the concentration of milk replacer or the offer of several litres of water in addition to milk replacer is recommended.

REFERENCES

Bogner, H., 1978. Einige Aspekte zur Haltung und Fütterung von Kälbern aus der Sicht des Tierschutzes. Berl. Münch. Tierärztl. Wschr. 91, 459-463

Bogner, H., 1980. Tierschutz in der Landwirtschaft - eine Herausforderung für die Ethologie. Der Förderungsdienst, Vol. 7.

Bogner, H. and Alps, H., 1980. Die Schätzung des Flächenbedarfs von Zucht - auf Mastkälbern auf Grund der Widerristhöhe. Unpublished.

Bogner, H., Schmitter, W., Schön, L., Ferstl, R., Scholz, W. and Peschke, W., 1973. Die Verwendung von Stroh als Einstreu und Futterergänzungsstoff in der Kälbermast. Bayer. Landw. Jahrb. 50, 90.

Bogner, H., Matzke, P., Ferstl, R., Alps, H., Seda, V. and Cisse, A.B., 1981. Strukturiertes Futter und Kälbermast. Tierärztliche Praxis, Jhrg. 9, Vol. 2, 181-187.

Gropp, J., Adam, G. and Boehncke, E., 1978. Der Natrium- und Kaliumgehalt von Milchaustauschfutter als Qualitätsmerkmal in der Kälbermast. Kraftfutter 11/78.

Groth, W., 1978. Tierschutz- und verhaltensbezogene Gesichts-punkte der Kälbermast. Der Tierzüchter, 30. Jhrg., Nr. 1-12.

Porzig, E., 1969. Das Verhalten landwirtschaftlicher Nutztiere. Deutscher Landwirtschaftsverlag, Berlin.

Van Putten, G. and Elshof, W.J., 1977. Zusatzfütterung von Stroh an Mastkälbern. Aktuelle Arbeiten zur artgemässen Tierhaltung. KTBL 233, S 210-219.

Reissig, F., 1979. Ruheverhalten und einige Aktivitäten von Mastkälbern in neuzeitlichen Haltungssystemen. Dissertation, Hohenheim.

PHYSIOLOGICAL ASPECTS OF SUCKING AND WEANING WHICH MAY AFFECT BEHAVIOURAL AND PATHOLOGICAL PROCESSES

F.R. Bell
Department of Medicine, Royal Veterinary College,
London, NW1 OTU, UK.

It is now well recognised that artiodactyls underwent an evolutionary surge in the Eocene Period because of the great development of the grasses which occurred at that time. The main kick to evolutionary development was the formation of diverticula in the fundic part of the stomach which permitted the fermentative digestion of cellulose by symbiotic cellulase producing bacteria. These large diverticula slowed up the passage of food through the alimentary tract and allowed direct absorption of short chain volatile acids from the rumen which yielded metabolisable energy. A further result was that carbohydrates were no longer absorbed as monosaccharides. The effect on the abomasum, the functional homologue of the simple stomach in the adult ruminant, was to relegate it to the role of a pump to transfer ruminal chyme to the small intestine for further digestion and absorption.

These excellent survival mechanisms of the adult ruminant artiodactyl, however, were of little value to the pre-ruminant calf which like other neo-natal mammals derived its nutritional elements from milk. The gastric diverticula are established during intra-uterine life between day 28 and 56 but they are still vestigial at birth when the abomasum has a volume equal to the other compartments of the stomach. This inhibition of the evolutionary phylogeny of the adult ruminant seen in ontogeny is further halted during immediate post-natal life. The specialised gastric diverticula of the adult stomach are of no value to the neo-natal calf which is sustained by milk which yields its energy by biochemical processes similar to simple stomach animals. In the calf this transitional period lasts about 3 weeks and is a time of rapid behavioural change, many aspects of which are brought about by equally rapid changes in metabolism (Bell, 1980).

Ruminants, like other animals, have a complicated series of correlated positive and negative processes between the brain and the alimentary tract which govern the behavioural mechanisms which together control food intake. A group of hormones evolved in the alimentary tract control motility, secretion, digestion and absorption by activating peripheral and central receptors to modify and integrate brain mechanisms to produce the specific feeding behaviour seen in different species. Metabolites also make an important contribution from peripheral receptors to the brain (Bell, 1976).

The calf, however, presents physiological, anatomical and metabolic problems in the post-natal period when the young ruminant derives its energy from milk and not from short chain volatile fatty acids. The cow provides very little nursing support for its calf nor is this necessary for the calf is capable of developing active, purposeful movements within a few minutes following parturition. Furthermore, temperature regulation and homeostatic mechanisms such as concentration of urine in the renal tubules by hormonal mechanisms are well developed at birth. Calves may be born with some degree of sodium deficiency but this is quickly moderated if they are allowed access to sodium salts.

The reticulo-rumen is vestigial in the calf compared to the adult bovine animal. The volume of the abomasum is about 1.5 l at birth and the non-functional reticulo-rumen has the same volume. The calf sucks only some 3 - 6 times/day and is dependent upon adequate stimulation of a neuro-endocrine reflex for milk 'let-down'. This process is important for survival especially in calves born to heifers or when other inputs from sore teats or painful foot lesions may inhibit contraction of mammary myoepitheliod cells as a result of adrenal catecholamine release.

In order to permit milk to pass directly to the abomasum the reticular groove is closed reflexly from pharyngeal receptors, activated dually by milk and the act of sucking. This reflex is very powerful and is very necessary to allow the colostrum to pass directly to the abomasum so that immunoglobulins can be transferred to the duodenum for absorption to establish the passive immunity necessary to combat pathogenic enteric organisms. Milk may be transferred to the reticulo-rumen when the pharyngeal receptors are not properly activated which often occurs when milk is drunk from a bucket. There is only a very slow transfer in the young calf of sequestered rumenal fluid to the abomasum, although reticular contractions are detectable even at a few days of age. It may take at least 4 - 5 days for milk to pass from the rumen to the abomasum.

When a calf is fed milk the reticular groove reflexes are reinforced and the characteristic behaviour of sucking and butting is maintained for many weeks. The infrequent drinking periods seen in calves are ideal for *ad libitum* feeding of groups of calves from a 'self-help' cafeteria. Bunching is avoided and each calf can feed without competition. This system of feeding has now almost replaced the single pen system and appears to be a more satisfactory procedure from the welfare point of view.

The calf is born with a drive to lick and nibble, a behaviour which is very advantageous for the ingestion of fomites containing the fauna and flora of rumenal substrate. This ingested material normally finds its way to the rumen but some long pieces of grass and straw and foreign bodies such as sheets of polythene often gain access to the abomasum. Since these substances are indigestible in the abomasum they can become compacted to obstruct the pyloric antrum with consequential pathologic change and death.

The abomasal mucosa in the newborn calf, like many other mammalian species, is almost free of parietal cells. This is advantageous in that the immunoglobulins of the colostrum are not degraded by proteolytic enzymes and pass directly in the abomasal chyme to the duodenum for absorption. In the calf some trophic factor activates an explosive proliferation of parietal cells within 12 - 24 h which stimulates active pepsin digestion in the abomasum. Calves which for one reason or another fail to ingest colostrum within this short period of time are denied the development of immunity needed to withstand pathogenic forms of enteric organisms. Bad management and weakly calves are the main cause of this situation.

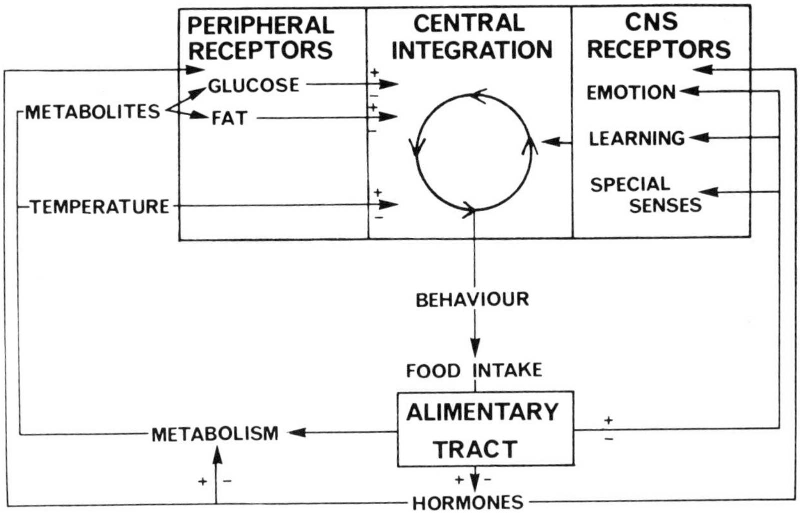

Fig. 1. Schema showing correlation of the control systems which regulate food intake. The central integration is probably brought about by the effects of neurotransmitters generated by a variety of sources on the metabolism of neurones. +, Stimulating effect; -, inhibitory effect.

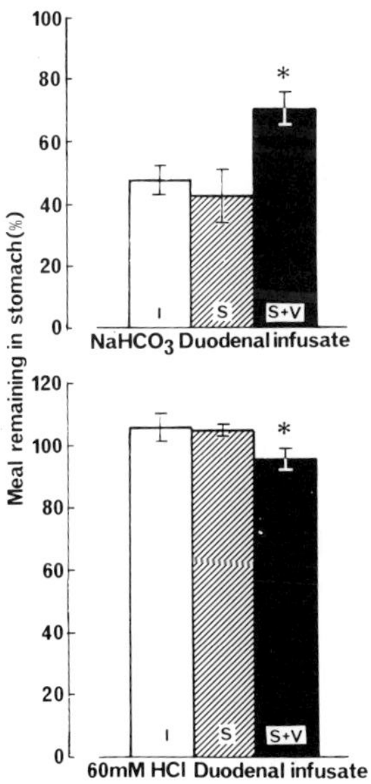

Fig. 2. The effect on gastric emptying of duodenal infusion with isotonic $NaHCO_3$ or 60 mM HCl in intact (I), splanchnectomised (S) and splanchnectomised + vagotomised animals (S+V). Significant differences in gastric emptying, v. I, are shown by *($P < 0.05$).

Once the milk is transferred to the abomasum a large clot is formed within a few minutes. The whey is separated rapidly from the casein clot so that lactose and whey proteins are transferred to the duodenum by the segmental contraction of the pyloric antrum. Earlier investigations in the milk-fed calf have shown unequivocally that it is the nature of the duodenal chyme which controls abomasal emptying (Figure 2). Alkalinisation of the duodenum activates abomasal emptying probably through reflex mechanisms originating from osmoreceptors in the small intestine. The intermittent periodic gushes of gastric chyme transferred to the duodenum depend upon the volume of liquid present in the body of the abomasum. The rate of antral systole is Ca. 3.5/min which delivers an intermittent small volume of abomasal chyme to the duodenum. During overfeeding, and the calf abomasum can readily be distended from 1 l to 4 l, the volume of the periodic antral gush increases so that the half-life of the abomasal contents is always the same (Ca. 25 - 35 min) whatever the initial volume (Bell et al., 1981). This increase in emptying also produces distension of the small intestine which precipitates mechanical diarrhoea which not only deprives the calf of calories but the loss of Na^+ reduces plasma osmolality which, together with the reduction of plasma pH to 7.0, causes fall in blood pressure, shock and anorexia.

The empty abomasum has a pH of < 2.0, a value which in the duodenum activates a system which inhibits abomasal emptying and gastric acid secretion. This inhibitory process is mediated by a number of hormones of which somatostatin is probably the most important but secretin and cholecystokinin are also involved. The calf appears to be unique in that this enterogastrone process is aided by unique morphological features in that the alkaline bile and pancreatic secretion are transferred to the duodenum through the hepatic duct 20 cm from the pylorus and 15 cm further caudally via the pancreatic duct. This anatomical arrangement would appear to maintain an acidic environment to enhance the elaboration of hormones which inhibit abomasal function. The inhibitory process is very powerful for volumes up to 4 l have been shown to be retained. It is obvious that excessive acidification of the duodenum could hold back gastric contents which on release to the small intestine could produce diarrhoea. (Figure 3).

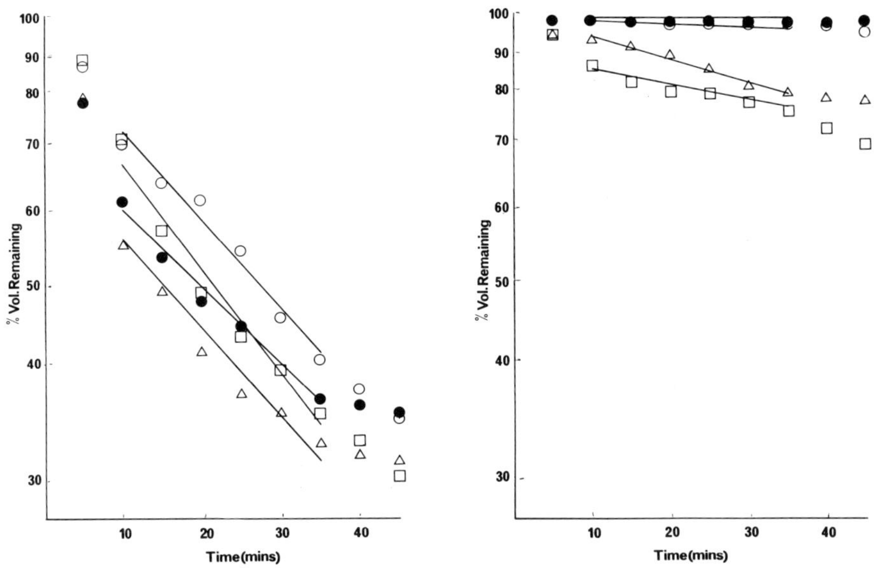

Fig. 3. The emptying of water test meals from the stomach (abomasum) of preruminant calves. The duodenum was under constant infusion with water at 10 ml/min at 39°C. The gastric output was collected directly from a proximal duodenal cannula for each 5 min period after instillation of test meal into the abomasum. The volume of meal instilled is shown: ●, 1.0 - 1.5 l (n = 22); O, 2 l (n = 4); Δ, 3 l (n = 8); □, 4 l (n = 4); n being the number of experiments used to compile the average data. The linear regressions drawn in for each volume of meal were calculated from ln of the mean of the percentage volume of test meal remaining in the abomasum at each 5 min period; data from the exponential phase of emptying (10 - 35 min) was used.

The inhibition of emptying water test meals from the stomach (abomasum) of preruminant calves. The conditions were the same as those mentioned above except that 0.06 M-HCl was used as duodenal infusate. The volume of meal instilled is shown: ●, 1.0 - 1.5 l (n = 19); O, 2 l (n = 4); Δ, 3 l (n = 7); □, 4 l (n = 4); n being the number of experiments used to compile the average data.

It has been suggested that a rise in PCO_2 in the duodenal lumen may be responsible for the inhibitory effects produced by acidification. Experimental examination shows that raised PCO_2 in the duodenal lumen is unimportant and that direct CO_2 stimulation of 'acid' receptors in the duodenum does not occur. Gases may occur in the abomasum, however, either from ingestion of air from a large artificial teat and bottle or from fermentation of material in the stomach including the rumen. Such gas in physiological amounts is normally voided by eructation from the normal gas cap but excess amounts produce abomasal tympany and inhibition of sucking. Usually the tympany is relieved by transfer of the gas to the intestine from where it is voided as flatus (Bell et al., 1981).

It has been suggested that increased acidification of milk replacers improves their nutritional value. This does not appear to be so since experimental investigation shows that gastric function is the same when the pH ranges from 2.0 to 12.0. The ingestion of a milk substitute by sucking from a bottle elevates the abomasal pH to 6.5, which during the process of digestion and transfer of gastric contents to the duodenum falls to the resting value (pH 1.5) in four hours (Webber et al., 1980). (Figure 5).

When the calf reaches the age of 21 days a crisis may occur, due to the fact that at this stage the abomasum may be subjected to overdistension. The calf is now beginning to take in reasonable quantities of concentrated food prior to weaning and often this accumulates in the abomasum. The calf continues to show very great motivation for the ingestion of milk or milk substitute and during the drinking process hypoxia may develop together with tachycardia and hypertension. The combination of reduced venous return and hypoxia may cause fatal convulsions due to cerebral anoxaemia. This situation can readily be avoided by management arrangements to prevent rapid ingestion of milk. The best method appears to be, to provide the liquid food from a self feeding system.

The effect of test meals of different temperatures on gastric emptying and secretion has been assessed experimentally. The temperature of the duodenal infusate does not affect gastric

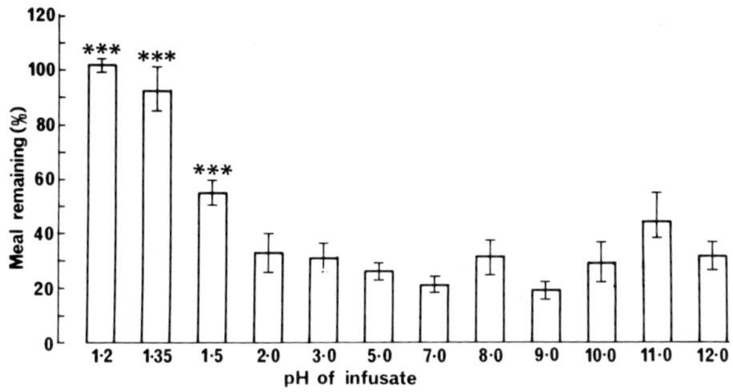

Fig. 4. Emptying of test meals, expressed as the mean percentage (± SE) remaining in the stomach after 45 min. The duodenum was perfused with isotonic NaCl solutions of pH 1.2 - 12.0; adjustment of pH was made with HCl or NaOH. Significant differences compared with the infusate at pH 7.0 are shown by ***($P < 0.001$). Each value is the mean of five experiments in four or five calves.

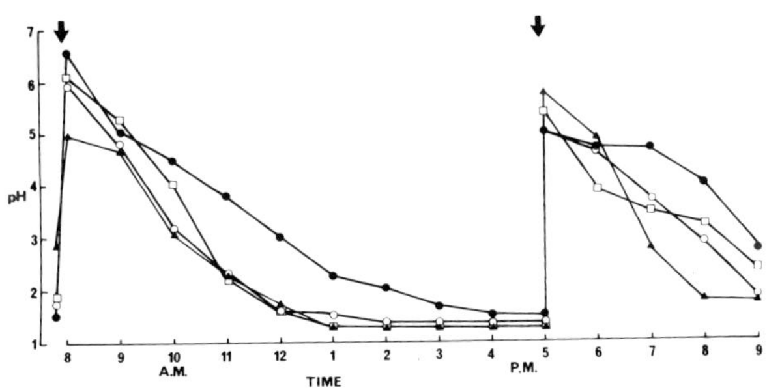

Fig. 5. The pH of intra-abomasal contents after sucking (arrows) 1.5 l milk substitute from a bottle (BOCM Silcock 176 Gold Top Baby Calf Food, 6.6% w/v, pH 6.6).

function. Variation in the temperature of the ingested meal results in a significant enhancement of gastric emptying at 4°C compared with 40°C. Shivering always occurred when 4°C meals were instilled into the stomach and intermittently with 20°C test meals, although gastric emptying is always improved to some degree by cold meals. The onset of shivering following the introduction of cold fluid into the stomach indicates that gastric thermoreceptors are concerned in the whole gamut of thermoregulatory control which is correlated with enhancement of gastric function (Webber et al., 1980) (Figure 6).

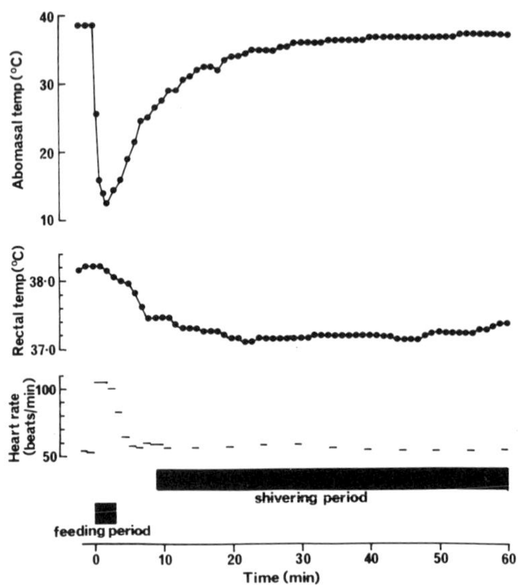

Fig. 6. Results from a single experiment showing the effect of consumption of 1.5 l cold milk replacer on intra-abomasal temperature, rectal temperature, heart rate and shivering.

One of the main problems of the neo-natal calf is the transition to the ruminant state. Calves show an ability to regurgitate even at 5 days of age. At 3 weeks of age calves graze for about 3 h a day and by 7 weeks have attained almost adult grazing times and adult blood VFA levels (Godfrey, 1961). The presence of VFAs in the rumen is a very great trophic stimulus to epithelial and muscular growth and may be the factor associated with digestion of cellulose which controls the duration of the period of dependence on milk. If this should be the case then the modern method of feeding calves by cafeteria and bedding them on straw may present a problem. The increased growth derived from cellulose digestion probably balances the nature of the protein laid down from milk feeding.

ACKNOWLEDGEMENT

This work was supported by a grant from the Agricultural Research Council.

REFERENCES

Bell, F.R., 1980. The mechanisms controlling abomasal emptying and secretion. In: Digestive Physiology and Metabolism in Ruminants. Eds: Ruckebusch, Y. and Thivend, P. p 81-100.

Bell, F.R., 1976. The regulation of food intake. Proc. Nutr. Soc., 35, 63-67.

Bell, F.R. and Watson, D.J., 1976. The influence of gastric distension and the duodenal infusate on the pattern of stomach (abomasal) emptying in the preruminant calf. J. Physiol., 259, 445-456.

Bell, F.R., Nouri, M. and Webber, D.E., 1981. The interplay between hydrogen ions, bicarbonate ions and osmolality in the anterior duodenum modulating gastric function in the conscious calf. J. Physiol., 314, 331-341.

Webber, D.E., Nouri, M. and Bell, F.R., 1980. A study of the effects of meal temperature on gastric function. Pflügers Arch., 384, 65-68.

Godfrey, N.W., 1961. The functional development of the calf. II. Development of rumen function in the calf. J. Agric. Sci., 57, 177-183.

DISCUSSION

N. Steenkamer *(Netherlands)* Calves have been fed cows milk containing casein for a very long time. More recently, cows milk in dairy breeds has been substituted by milk replacer containing, up to a few years ago, casein as well in the form of skimmed milk powder. During the last few years several zero milk replacers have been developed containing skimmed milk powder and not casein. Does this change your story of emptying the abomasum, the effect of management errors such as overfeeding? If this is true and it is not easy to prove, then by feeding zero milk replacers, if they are the right composition and quality, then the incidence of scour will be lower than with the skimmed milk powder formula, and the incidence of bloat will also be lower. I have never really heard of a good explanation of this phenomenon, but let us suppose that this phenomenon is true.

F.R. Bell *(UK)* I do not know whether what you say about the zero protein is correct. I think what you mean is that the casein is replaced by soya, a vegetable protein.

N. Steenkamer The casein is replaced by whey proteins, delactose whey powder in the ration, part will be casein as such, sodium caseinate, and part will be soya products, concentrates and sometimes isolates.

F.R. Bell In our experience, milk replacers are very good. However, other people like Unilever, and I am sure you have read the report by Porter, contend that this affects the villi of the small intestine. These villi are reduced to a very small size which can affect the absorbative properties, possibly the production of the hormones I have mentioned and it may have some effect in the absorption and exsorption process which goes on across the duodenum mucosa.

I cannot answer your question, all I can say is that Unilever contend that this has a pathological effect.

N. Steenkamer As to the emptying, it has been shown that the emptying of the abomasum is faster and there is no clotting in the abomasum, which is easily understood. Furthermore, soyas are not all the same. There are soya products containing some peptides responsible for changing the brushborder. Other soya protein preparations do not seem to have this property. This has been shown in piglets, and in Holland there is an interesting development where the small intestines of piglets were removed from the animals and examined for the presentation of the villi and the brushborder. Different reactions to various soya preparations were observed. The results have not yet been published but I would say that various types of soyas might give this phenomenon others might not, at least not to the same extent.

F.R. Bell The factor which controls emptying is the isotonicity through osmoreceptors. They can be activated in a whole variety of ways and I just don't know how these other replacers will affect it. But you have seen that isotonicity is the thing which drives. The normal state for an abomasum is to be emptying and it is only inhibited occasionally because the whey is always being produced and the protein is always there.

J.F. Grongnet *(France)* Would you please repeat your opinion about feeding colostrum in relation to the production of hydrochloric acid by the abomasum.

F.R. Bell The calf is distinctive from other species of animal, or, to put it more widely, the young ruminant artiodactyl - lambs, calves, kidlets and the young of the wildebeest and the yak - all these animals that breed so successfully. All these animals are born with very few parietal cells. Now these parietal cells are developed very rapidly and are there in adult numbers in 24 h. There is something which stimulates them or inhibits them. As a result of that, the immunoglobulins which are produced from the whey and passed to the duodenum for absorption can then transfer their precursor of the immunoglobulin of the plasma in that period of time. If the calf does not take in the immunoglobulins in that period of time which may last from 6 - 12 h, then when it does get the colostrum at that time it will digest it. If it has digested it, then it has of course changed its form and is absorbed in another way. It will provide nutrients but not immune body.

J.F. Grongnet In your opinion then it is not of real use to give colostrum after 12 h of life?

F.R. Bell That is the normal practice.

D.B. Stephens *(UK)* Do you think that if we offered supplementary colostrum to the calf early on after birth that this might inhibit or diminish the sucking drive in any way?

F.R. Bell I have no experience of that. I would advocate that every calf is given colostrum at the earliest opportunity either by stomach tube, by sucking or preferably from its own dam. That is a management decision. But if they are deprived of it, then they will be susceptible to all forms of bacterial invasion.

D.B. Stephens What I meant was that if you had a sort of colostrum bank, you could then offer say a couple of litres to each calf.

F.R. Bell Yes, that is admirable and, I think, common practice, certainly in the south of England.

D.B. Stephens But do you think that it will affect the sucking drive?

F.R. Bell No, I would not think so although I have no proof of that.

A.J. Webster *(UK)* Was Dr. Steenkamer talking about milkless milk replacers which are also acidified or not?

N. Steenkamer It does not matter whether they are acidified or not. The phenomenon is exactly the same with acid added or not - the reduction in scour and bloat. We have never seen any difference. The acid is added for publicity and commercial reasons but not, and I agree fully with what Dr. Bell said, to reduce incidence of scour by acid as such.

J.Y. Blom *(Denmark)* Professor Bell, we do not disagree on the subject of early colostrum intake but I do have two brief questions. The first is that you said in your paper that the triggering factor of the development of the acidic cells, the parietal cells, is in fact the casein.

F.R. Bell No, in fact I was careful not to say that. I said it might be a stimulant or it might be inhibitory.

J.Y. Blom If you delay the intake of colostrum, do you also delay the development of the parietal cells?

F.R. Bell As far as I know, that does not happen.

J.Y. Blom To support Dr. Stephens, I can say that Dr. Stott in America and his co-workers, especially Nightingale, have found in a cattle herd that if they gave an extra supplement of colostrum after the suckling period, that is, within the first 24 h (say, the calf was left overnight with the mother in the calving box and then given a supplement of colostral feeding) then they found that the antibody titre in the calf serum did in fact rise. How does this fit in with your ideas?

F.R. Bell I don't know the answer to that one either. Very few people would agree that after 24 h that the titre would go up, but it may. I have been specific on 6 and 12 h but like every biological thing you must take it with a grain of salt. It might be 14 h or 4 h. It is very difficult to get sufficient statistics to be absolutely precise.

THE INFLUENCE OF HOUSING AND CLIMATISATION ON HEALTH AND GROWTH OF YOUNG CALVES UNDER FARM CONDITIONS

J.Y. Blom
Institute of Internal Medicine,
Royal Veterinary and Agricultural University,
Copenhagen, Denmark.

INTRODUCTION

The recent changes in the pattern of cattle farming in Denmark, as in other industrialised countries, have given rise to severe disease problems. The calves encounter these disease problems most especially in specialised calf fattening units, but also in dairy herds. These problems are emphasised by a yearly calf mortality rate of approximately 15 - 18% in Danish cattle herds. There has been a steady decrease in the number of herds, while at the same time herd sizes have increased considerably. The manpower used to keep the animals is decreasing due to high labour costs.

The altered production patterns outlined above are important in understanding most calf diseases as they are typical examples of multifactorial diseases, i.e. diseases that develop as a result of the combined influences of primary disease agents (viruses, mycoplasmas, bacteria) and adverse factors in environment and management.

The prevailing disease situation in calves in Danish dairy herds was recently studied in order to investigate the influence of housing and climatisation on the health and growth of young dairy calves (Blom and Thysen, 1980; Henneberg and Thysen, 1980; Blom, 1981).

MATERIALS AND METHODS

A total of 1 038 calves in eight dairy herds were studied from December 1st, 1978 to May 31st, 1980. The calves were fed colostrum at birth and transferred to the calf house within two days *post partum* where they remained for approximately four months. All calf houses were isolated, and ventilated by pressure ventilation (fresh air distributed from a tapered inlet duct at the ceiling centreline). In three of the houses the inlet air could be heated from the start of the investigation. Heating was installed in all houses by the end of the first winter period due to severe health problems.

The calves were placed in individual pens for the first eight weeks, after which they were transferred to group pens for the remaining two months in the house. All calves were fed standardised amounts of milk substitutes and were offered concentrates and hay from the second week of life.

The climate in the houses was recorded by thermohygrographs during the entire study period. The weights of the calves were recorded at birth and thereafter at monthly intervals. All veterinary treatments of sick calves were recorded and all data were transferred to computer files for processing.

RESULTS

As can be seen from Table 1, respiratory disease (RD) was the major disease problem in all herds. The overall incidence rate for RD was 29%, but varied amongst herds from 14 - 43%.

TABLE 1

INCIDENCE RATES FOR THE VARIOUS DISEASES IN THE INDIVIDUAL HERDS

Herd	No. of calves	Respiratory disease %	Enteritis %	Omphalitis %	Other diseases %
A	120	14	3	5	3
B	78	36	3	0	1
C	90	21	7	1	2
D	149	24	3	0	2
E	150	15	5	1	1
F	192	38	9	5	1
G	112	43	16	2	5
H	147	38	1	0	7
Total	1 038	29	6	2	3

Mortality rates in the herds are shown in Table 2. The overall mortality rate was 6.8% (day 2 - 120), but varied amongst herds from 0.7 - 15.2%. Respiratory disease was the

TABLE 2

MORTALITY RATES IN THE INDIVIDUAL HERDS, DAY 2 - 120.

Herd	Total mortality %	Mortality due to respiratory disease %	Mortality due to other causes %
A	7.5	3.3	4.2
B	2.6	1.3	1.3
C	2.2	0	2.2
D	6.0	5.4	0.6
E	7.3	4.0	3.3
F	10.4	7.8	2.6
G	15.2	11.6	3.6
H	0.7	0	0.7
Total	6.8	4.5	2.3

major cause of deaths (66%), but this was due mainly to the high incidence rate.

No relationship was found between parity of the dam and the incidence of RD, enteritis and mortality rates, and the incidence of disease and the mortality rate was not related to the sex of the calves.

The relationship between the age of the calves and the incidence rate for RD is shown in Figure 1. There was a slight peak of incidence in weeks 5 - 7. However, the age of the first case of RD was highly dependent upon the total disease situation in the individual herds, as can be seen from Figure 2.

The data presented in Table 1 suggested a relationship between the occurrence of RD and other diseases in the individual herds. This relationship was further analysed, and the results are shown in Table 3.

The seasonal distribution of RD in the herds is shown in Figure 3 using a 'cusum' technique. The cumulative total of new courses of RD (i.e. the incidence) was thus compared to the expected incidence, based on the calf population present and the age structure of the calf population at a given time. As can be seen from the Figure, the incidence of RD coincides with the calving season in most herds. Therefore, the influence

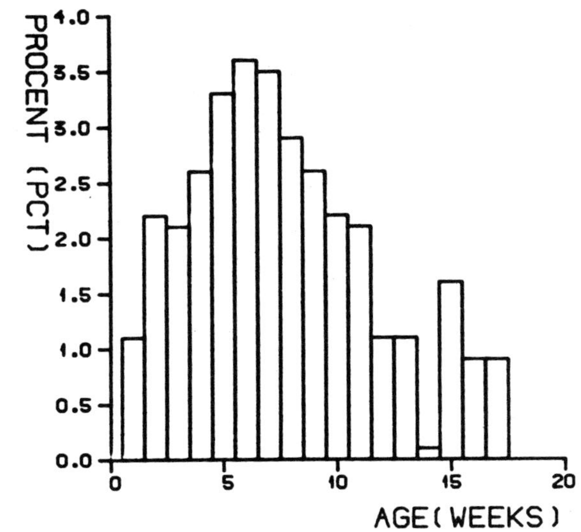

Fig. 1. Incidence of respiratory disease in relation to age.

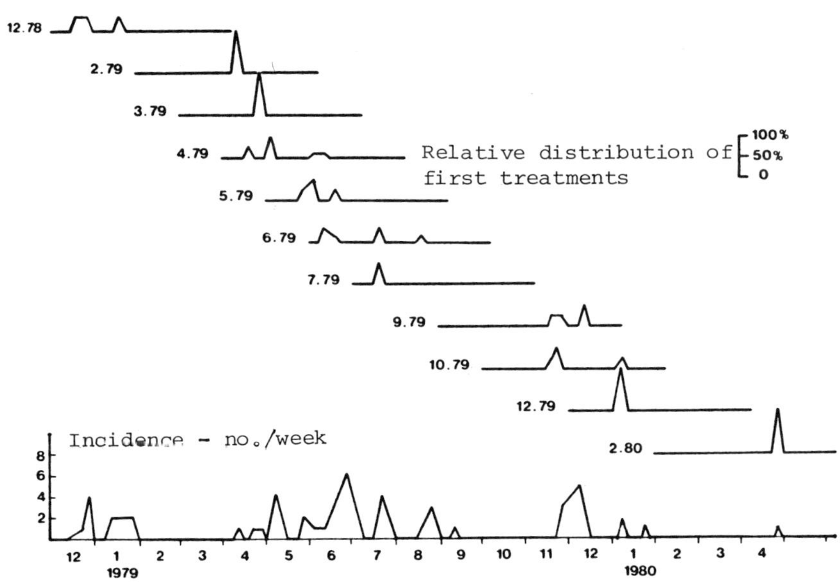

Fig. 2. Age at first case of respiratory disease in relation to month of birth (upper curves) and occurrence of respiratory disease in the herd (bottom curve). Herd D.

TABLE 3

THE RELATIONSHIP BETWEEN NEONATAL DISEASE[1] AND THE INCIDENCE RATE AND CASE FATALITY RATE FOR RESPIRATORY DISEASE.

Disease category	Incidence No.	Incidence Population at risk	Incidence %	Case fatality No.	Case fatality %
Neonatal disease, total	81	1 038	7.8	15	18.5
Neonatal disease and respiratory disease	30	66	45.5**	8	26.7*
Respiratory disease without previous disease	268	957	28.0	31	11.6

[1]Enteritis, omphalitis, arthritis, generalised infections.

$**x^2 = 9.11$, $p < 0.01$ $*x^2 = 5.05$, $p < 0.05$

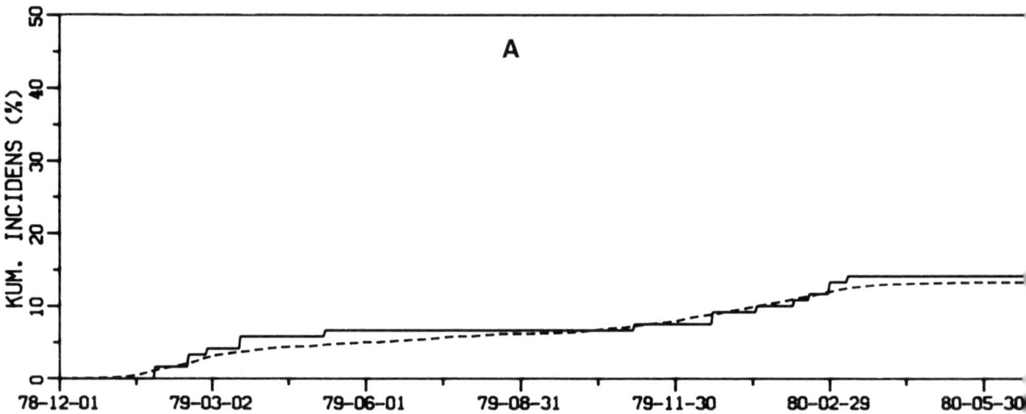

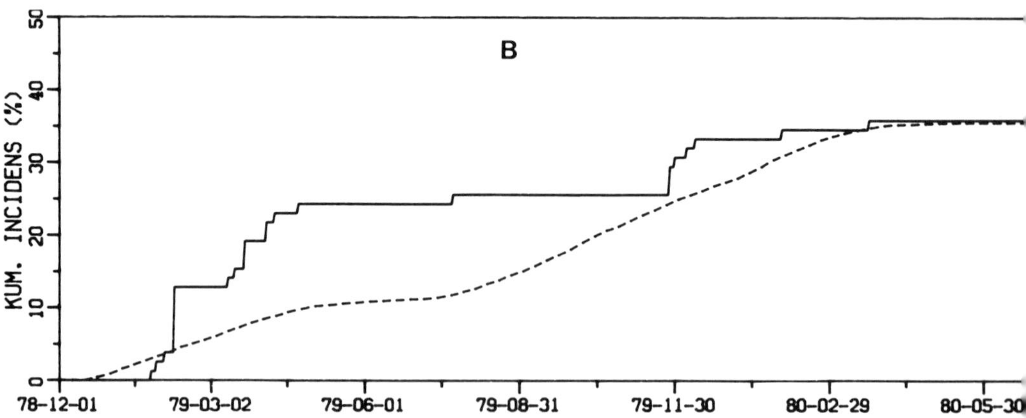

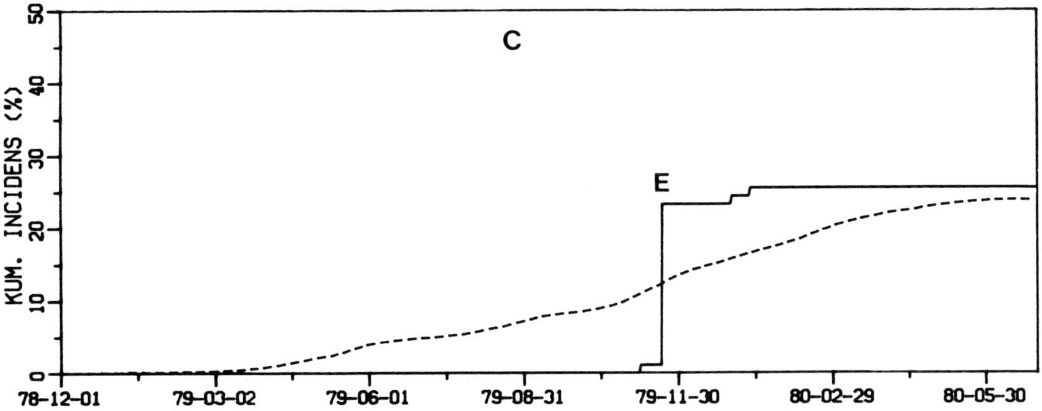
C
E
KUM. INCIDENS (%)
0
10
20
30
40
50
78-12-01
79-03-02
79-06-01
79-08-31
79-11-30
80-02-29
80-05-30

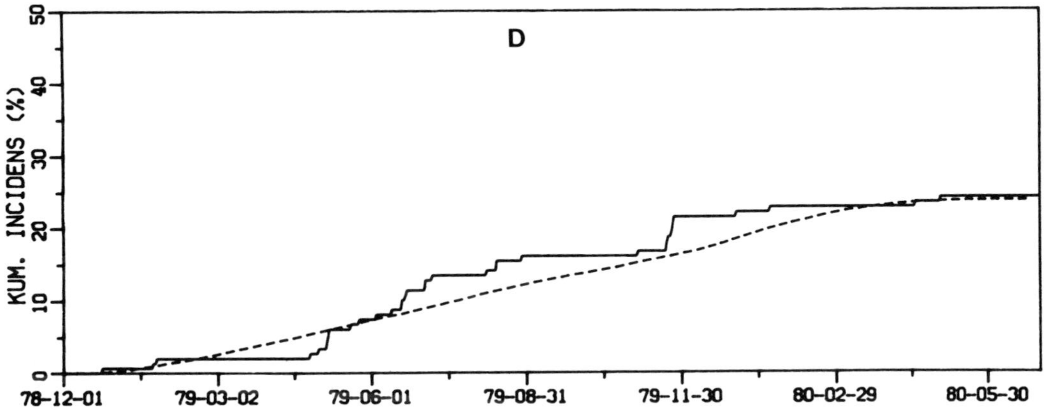
D
KUM. INCIDENS (%)
0
10
20
30
40
50
78-12-01
79-03-02
79-06-01
79-08-31
79-11-30
80-02-29
80-05-30

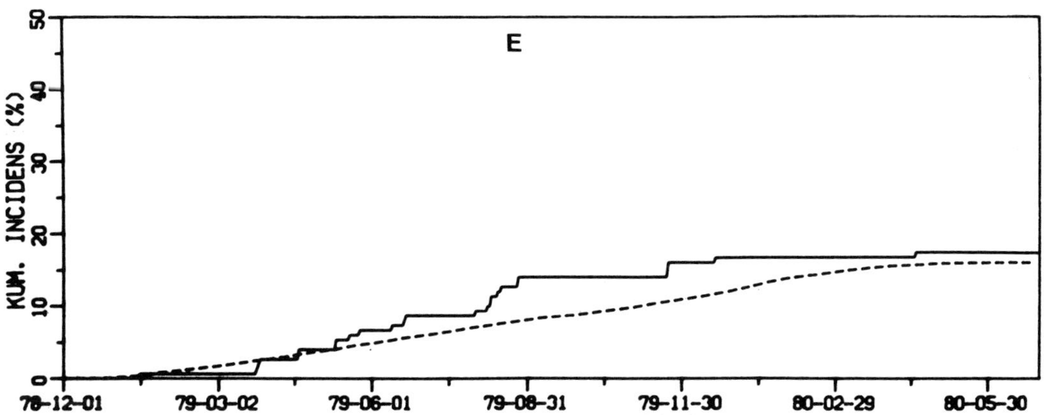
E
KUM. INCIDENS (%)
0
10
20
30
40
50
78-12-01
79-03-02
79-06-01
79-08-31
79-11-30
80-02-29
80-05-30

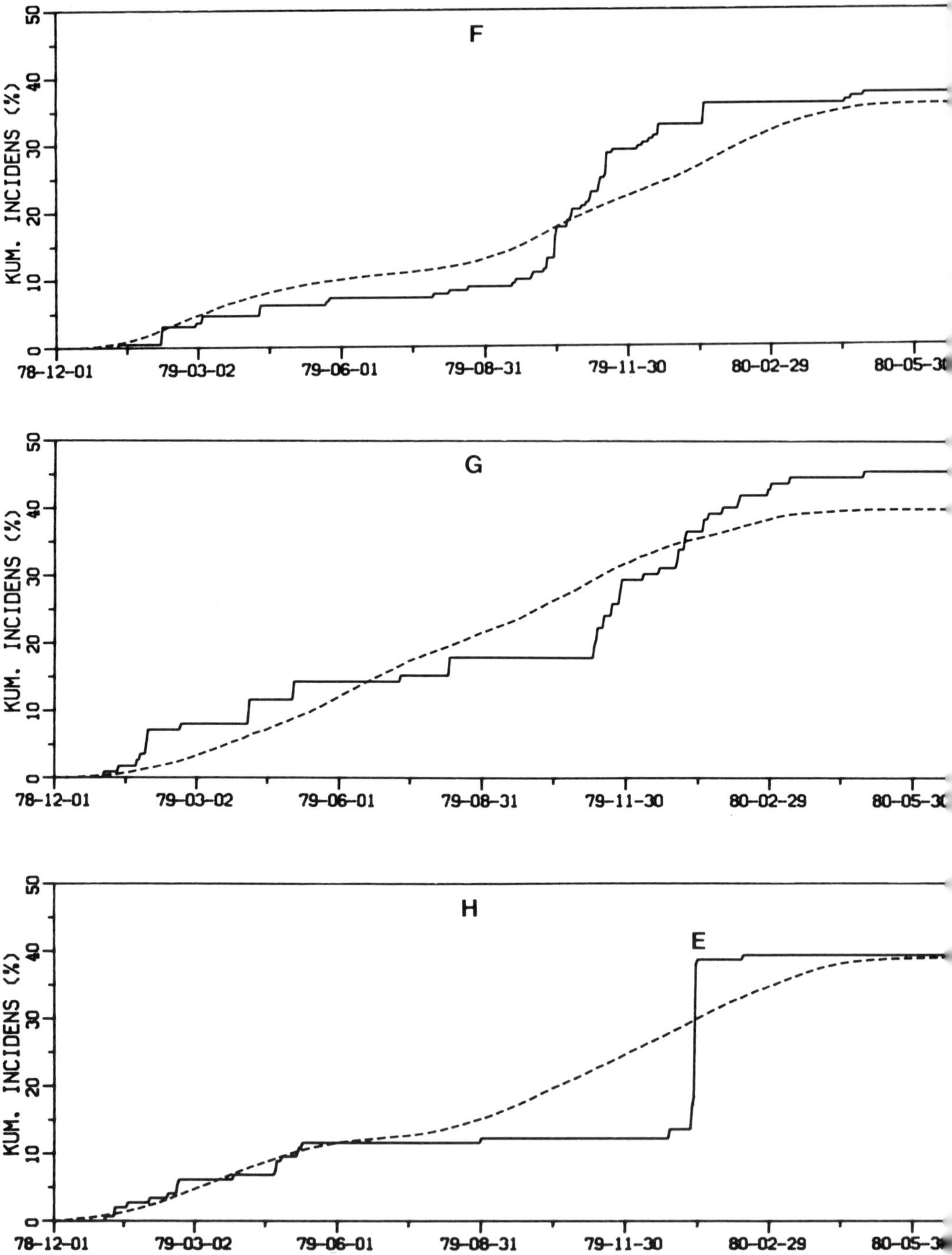

Fig. 3. Cumulative incidence of respiratory disease in the individual herds throughout the study period, observed (——) versus expected (----) distribution. 'E' = major epidemic.

of climate in autumn and winter can be difficult to separate from the effect of a high stocking rate in these periods. In herds D and E, calvings were evenly distributed throughout the year and it is noticeable that a high incidence of RD was found during the summer months in these herds.

To elucidate further the effect of climate on the incidence of RD a data analysis was performed for the whole study period and selected periods.

Table 4 shows the relationship between the incidence rate for RD and selected indices of climate and environmental factors. Herds with a high incidence rate of RD (>25%) were compared to herds with a low incidence rate. The analysis revealed that high humidity and sparse bedding were of importance.

TABLE 4

INCIDENCE OF RESPIRATORY DISEASE IN RELATION TO BARN CLIMATE, LOCAL ENVIRONMENT AND OTHER DISEASES[1].

Herd	Climate: No. days with >85% rH[2]	Climate: No. days with <10°C	Local environment: Bedding, kg/calf per day	% unclean calves	Other diseases %	Incidence rate for respiratory disease %
A	15	9	1.3	0	10	14
B	35	19	0.7	5	7	15
C	32	12	0.7	8	8	21
D	22	22	0.7	5	5	24
	$\bar{x}$=26 SD=9.2	$\bar{y}$=16 SD=6.0				
E	64	10	1.0	11	3	36
F	48	24	0.3	5	14	38
G	52	9	0.8	5	1	38
H	24	31	0.4	16	20	43
	$\bar{x}$=47 SD=16.8	$\bar{y}$=19 SD=10.8				

[1]Definitions given in Table 3

[2]rH = Relative humidity.

The effect of heating is shown in Table 5. The data in this Table were analysed by a weighted regression analysis, applying the logit transformation to the linear regression model (Grizzle et al., 1969; Snedecor and Cochran, 1967). The model used included the effect of heating and occurrence of predisposing diseases. The results of the analysis are given in Table 6.

TABLE 5

BARN CLIMATE AND INCIDENCE RATE FOR RESPIRATORY DISEASE IN RELATION TO USE OF HEATING[1]

Herd	Year	Heat-ing	No. days with >85% rH	No. days with <10°C	Incidence rate for respiratory disease %
A	1978	+	8	1	9.7
	1979	+	6	0	10.6
C	1978	+	1	10	14.3
	1979	+	17	0	35.7
F	1978	+	41	13	18.9
	1979	+	6	1.	46.0
B	1978	-	53	9	34.0
	1979	+	11	0	12.2
D	1978	-	22	15	24.7
	1979	+	0	0	13.6
E	1978	-	35	14	21.3
	1979	+	0	0	6.3
G	1978	-	46	24	42.3
	1979	+	0	2	33.8
H	1978	-	38	9	18.3
	1979	+	0	0	3.5

[1]Periods December - January 1978-79 and 1979-80.

TABLE 6

THE EFFECT OF HEATING AND 'OTHER DISEASES' ON THE INCIDENCE OF RESPIRATORY DISEASE, USING A WEIGHTED REGRESSION ANALYSIS ON THE DATA PRESENTED IN TABLE 5.

Variable	Estimate	x^2	Degree of freedom
Intercept	1.2621	94.79	1
Heating	0.8533	26.55***	1
Other diseases	-1.3137	66.29***	1
Residual effect		4.66*	1

*$p < 0.05$ ***$p < 0.0001$

Figure 4 and Table 7 show the effect of RD upon the growth rate of the calves. As can be seen from Figure 4, calves experiencing uncomplicated RD without relapses at an early age suffered no weight loss. On the other hand, calves that had RD at an older age of 71 - 98 days *post partum*, or had relapses, had a lower growth rate than healthy calves.

TABLE 7

AVERAGE DAILY WEIGHT GAINS FOR TREATED VERSUS NON-TREATED CALVES (HENNEBERG AND THYSEN, 1980)

	Average daily weight gain	
	At 130 days *post partum*	At 12 months *post partum*
Heifer calves		
Non-treated	638 g	-
Treated, %	94%	-
Bull calves		
Non-treated	765 g	978 g
Treated, %	86%	96%

Table 7 shows the relative growth rate for treated calves versus untreated calves. The average daily growth rate for treated heifer calves was 6% lower, and for treated bull calves 14% lower at 130 days. The difference in daily weight gain at slaughter for treated bull calves was 4%.

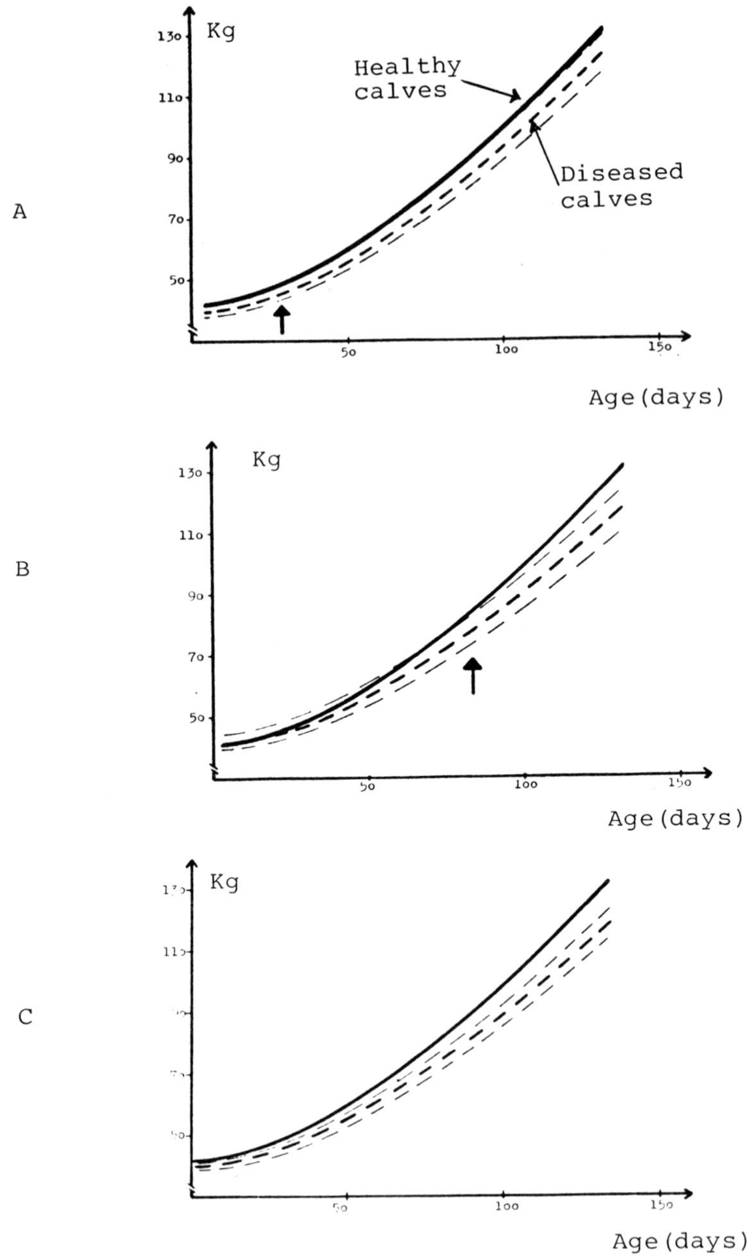

Fig. 4. Growth curves for calves. A: Calves with RD 15 - 42 days old. B: Calves with RD 71 - 98 days old. C: Calves with several disease incidents. Thin dotted lines indicate confidence limits ($p < 0.05$) (Henneberg and Thysen, 1980).

DISCUSSION

In this study respiratory disease was found to be the major health problem in all calf herds. This is in accordance with other reports (Jørgensen, 1973; Pritchard, 1980).

The data demonstrate that this disease complex is very costly to Danish cattle production.

The study revealed the age-dependent peak of treatments for RD at the age of 5 - 7 weeks that was also found by other workers (Miller et al., 1978), but it is interesting to notice that the age at first treatment is highly dependent on the overall disease situation in the herd, i.e. in herds with a continuous re-stocking of calves. The younger susceptible subjects are highly exposed to infection from older diseased calves.

The observed relationship between other diseases and RD is interesting as it reflects to some extent the management of the young calf. Some authors have found a direct relationship between the immunoglobulin status of the young calf and the incidence of RD, while others fail to demonstrate this relationship (Williams et al., 1975; Barber, 1978).

But as most calves suffering from neonatal diseases are immunoglobulin deficient (Blom, 1981, to be published), the indirect relationship between immunoglobulin status and RD can be explained from the results as shown in Table 3.

To elucidate the influence of the climate, an attempt was made to compare daily observations of climate with the incidence of disease without conclusive results. This may be explained by the fact that the two variables may not be coincident in real terms (Harkness, 1977). The 'cusum' technique applied in Figure 3 shows that the seasonal pattern of RD is often associated with a high stocking rate, but disease 'excesses' and major epidemics often appeared in the late autumn and winter periods when major weather changes occurred, followed by high humidity and major temperature fluctuation in the calf houses.

The analysis of the total data material, as shown in Table 4, revealed a relationship between respiratory disease and high humidity and sparse and therefore wet bedding in the calf boxes, whilst the health of the calves was not affected by differences in temperature.

Heating was installed in all houses by the end of the first winter period and it was possible, therefore, to compare two winter periods and so study the effect of heating. The data in Tables 5 and 6 show that heating did improve the health situation in almost all herds, and this improvement seems to be caused mainly by a drier climate in the houses which is in accordance with other reports (Harkness, 1977; Webster, 1979), although the causal relationship still remains unclear.

The daily weight gains of the calves were shown to be highly dependent upon the time and course of RD. Young calves, fully recovered from the first disease occurrence did not suffer from any weight loss, whilst calves that contracted RD later in life or had several relapses were shown to suffer a considerable loss in weight gain.

However, this loss was reversible as can be seen in Table 7. It is noticeable that treated heifer calves showed smaller weight losses compared to bull calves. This may be explained by the fact that healthy bull calves have a higher daily feed intake and weight gain than heifer calves. Therefore, disease is more costly to the production of fattening bulls, although the data indicate compensatory growth later in the life of surviving animals.

Veterinary research into calf diseases has dealt traditionally with the identification and study of the infectious organisms associated with disease in order to achieve disease control. As most of these diseases are multifactorial, epidemiological studies seem to be more relevant for the immediate control of calf diseases and further epidemiological studies should be undertaken to ensure healthy calf production.

REFERENCES

Barber, D.M., 1978. Serum immune globulin status of purchased calves: An unreliable guide to viability and performance. Vet. Rec., 102, 418-420.

Blom, J.Y., 1981. Enzootisk pneumoni hos kalve. Epidemiologi og profylakse. (Enzootic pneumonia in calves. Epidemiology and prophylaxis). Ph.D.-Thesis, Copenhagen.

Blom, J.Y. and Thysen, I., 1980. Klimaets indflydelse på kalves sundhed (The influence of climate on calf health). 502. Beretning fra Statens Husdyrbrugsforsøg, Copenhagen, p. 138-153.

Grizzle, J.E., Starmer, C.F. and Koch, G.G., 1969 Analysis of categorical data by linear models. Biometrics, 25, 489-504.

Harkness, J.W., 1977. Environmental studies and enzootic pneumonia of calves. Bull. Off. int. Epiz., 88, 3-15.

Henneberg, U. and Thysen, I., 1980. Sundhedens indflydelse på kalves tilvækst (The influence of the health status upon the weight gain of calves). 502. Beretning fra Statens Husdyrbrugsforsøg, Copenhagen, p. 154-166.

Jørgensen, M., 1973. Sygdom-miljø registrering. 410. Beretning fra Statens Husdyrbrugsforsøg. Copenhagen, 130-140.

Miller, W.M., Harkness, J.W., Richards, M.S. and Pritchard, D.G., 1978. Studies of calf respiratory disease in a large commercial veal unit. In: Martin, W.B.: Respiratory diseases in cattle. M. Nijhoff, The Hague, 181-195.

Pritchard, D.G., 1980. Current research on calf pneumonia. In: Grunsell, C.S.G. and Hill, F.W.G.: Veterinary Annual vol. 20, Scientechnica, Bristol, 189-203.

Snedecor, G.W. and Cochran, W.G., 1967. Statistical methods, 6. ed., Iowa State University Press, Ames.

Webster, A.J.F., 1979. Housing and husbandry of the veal calf. I-:Grunsell, C.S.G. and Hill, F.W.G.: Veterinary Annual vol. 19, Scientechnica, Bristol, 49-53.

Williams, M.R., Spooner, R.L. and Thomas, L.H., 1975. Quantitative studies on bovine immunoglobulins. Vet. Rec., 96, 81-84.

DISCUSSION

A.A. Jongebreur *(Netherlands)* Was the mortality rate that you mentioned not inclusive of stillbirths?

J.Y. Blom *(Denmark)* No, stillbirths were not included in the mortality rate.

N. Steenkamer *(Netherlands)* What do you consider in this context to be neonatal diseases - those in the first 24 h or during the first week? You did not specify this.

J.Y. Blom In this investigation we defined neonatal diseases as all those diseases in the first week of life.

D.B. Stephens *(UK)* In the last slide but one (Figure 3), I did not quite understand whether there was or was not an effect if they got respiratory disease on say day 19?

J.Y. Blom There was a significant drop in weight gain.

F.J. Harte *(Ireland)* In your last slide (Figure 4) you showed that your diseased cattle had recovered to within 4% of the liveweight of the control group one year after *post partum*. Were there any further deaths in the diseased group between the calf period and the final weighings because you will appreciate that it is usually the animals that are performing most poorly that die? This, in turn, tends to give the impression of a better performance by the diseased group in total than if they all remained alive.

J.Y. Blom Yes, I must admit that calves that died were not included in this analysis. We counted the weights of the dead calves. The Figure shows only the calves that were alive at the end of the analysis.

ENVIRONMENTAL AND CLIMATIC INFLUENCES ON PNEUMONIA OF CALVES

C. Holzhauer
Gezondheidsdienst Voor Dieren in Gelderland,
PO Box 10, 6880 BD Velt, The Netherlands.

In the Netherlands cattle farming is a major agricultural sector, comprising dairy farms, bull fattening farms, calf fattening farms and calf rearing farms.

On all these farms the rearing of young cattle is the basis of good management. In rearing calves stock farmers and veterinary surgeons are faced with various kinds of health disorders. The results are a temporary slowdown in growth, but also loss of animals due to mortality or permanent unfitness for the purpose for which they were intended. Moreover, there is the additional cost of treatment and medicaments.

In order to prevent or control these diseases attending calf rearing, we first of all need knowledge of, and insight into, the aetiological factors. Infectious diseases of the respiratory tract are a major risk in calf rearing and mostly occur in groups. About 50% of losses (mortality and animals culled) are caused by pneumonia.

The causes of infectious diseases of the respiratory tract are complex due not only to the presence of a pathogen, but also other factors, such as the environment. Accordingly prevention and therapy should not be exclusively based on control of the pathogen.

Various syndromes can be distinguished in infectious diseases of the respiratory tract in young cattle. Firstly, there is enzootic calf pneumonia which we observe in calves of approximately 30 - 60 days old. It is also called crowding disease, or calf influenza, and is identical with shipping fever. Calf pneumonia occurs on all the above-mentioned types of cattle farms and is by far the most important pulmonary disease, in that it has the highest incidence and the greatest economic significance. Secondly there is pinkengriep - or influenza in one year old heifers. It is an acute infection of the respiratory tract occurring in herds of older calves. Like calf pneumonia the course and end of the disease depend greatly on environmental factors. Thirdly there is lungworm, which has a parasitic aetiology and is thus tied to the grazing period. Fourthly, there is IBR and the course of this disease is also highly dependent on the environment.

In all these pulmonary diseases with an infectious aetiology the climate plays a vital role. Thus the development of the parasitic lungworm larvae depends entirely on humidity and temperature since these determine the time the larvae need for development and their chances of.survival, and therefore the extent of the infection. In calf pneumonia and influenza of yearling heifers the climate will usually be that in the calfhouse, the internal climate. We will now take a closer look at calf pneumonia from this angle.

The pathogenesis of this multifactorial disease shows primarily a viral infection of the respiratory tract, attended with catarrhal symptoms such as slight serous nasal discharge, conjunctivitis and sometimes a minor flow of saliva. There is coughing, and respiration is usually accelerated and changed. The general state of health is frequently disturbed; the animals are less active and may have a fever.

Sometimes the viral infection is localised in the front part of the respiratory tract, but in many cases will also affect deeper situated tissues (coughing). This infection and the consequent bacterial infection may be over in a few days, but may also spread. The course of the disease depends on the body's defence reactions and the extent or progress of the initial infection. Both infection and defence are connected with the climate. In the case of aerogenous infection the infectious agent will be transmitted via droplets of moisture formed in the respiratory tract of animals producing the infectious agent. The droplets containing the infectious agent float in the air and move along with the airstream. Part of them will sediment or leave the cowhouse through the ventilation system. The continued existence of the remaining droplets will depend on the relative humidity. In a dry climate they will evaporate more quickly than in a more humid climate. Thus the degree to which herdmates are infected will be related to the movement of the air, relative humidity, temperature and ventilation. In addition to the effect of climatic factors on the transmission of the infection, there may also be a direct influence on the infectious agent in the environment. As I have said before, there is another way in which the climate in the house affects the course of infections that occur, namely through defence factors. The effectiveness or ineffectiveness of the defence factors have been assumed to determine the progress of an infection, and hence the development of disease symptoms and their extent and duration. Both phagocytosis and immune response are affected by the environment, the house climate and, directly or indirectly, the outdoor climate again playing a significant role, besides other housing factors, such as bedding, number of animals per pen, and social peace.

In Germany, Andreae et al. (1968) have described a relation between outdoor climate and pulmonary disorders in young cattle. In the winter months many treatments for pulmonary disorders were necessary in a group of some 1 200 calves after the occurrence of depressions. From this the authors feel justified in concluding that a relationship exists between these depressions and the genesis of infections of the respiratory tract. This connection between changes in the weather and pulmonary disorders has also been observed elsewhere. We ourselves have had the opportunity to make some observations regarding the relationship between house climate and infections of the respiratory tract on the regional experimental farm 'De Vlierd'.

For eight years the incidence of calf pneumonia in an insulated rearing house was compared with that in an uninsulated rearing house. As a rule 80 calves from one week to three or four months old were kept in littered group pens, 40 in the insulated and 40 in the uninsulated section. Both sections were mechanically ventilated and temperature and humidity were recorded in both. Each year there were more sick animals in both houses in November than in February and May, and the number of cases in February was higher than that in May. The percentage of sick animals was between 5 and 20, and once as high as 50. It was remarkable, and contrary to expectation, that there were always more sick animals in the insulated section than in the uninsulated section.

What were the climatic differences between these two sections? In the uninsulated house there were large fluctuations in temperature and relative humidity of more than 10°C and 20% RH/day. In the insulated house the fluctuation in temperature per day was only a few degrees, and the relative humidity varied little and was constantly high (> 90). After five years it was decided to try to bring the relative humidity in the insulated house as much into line as possible with that in the uninsulated house by heating the incoming air when the relative humidity had become too high. These efforts were reasonably successful.

The difference in the incidence of disease caused by infections of the respiratory tract continued to exist with more sick animals in the insulated house. We concluded that fluctuations in temperature and relative humidity are not the only climatic factors involved in the genesis of lung disorders in groups of calves.

The third point on which the houses differed - the ventilation and ventilation pattern - might be a more important factor. On commercial farms it is often possible to reduce the excessively high incidence of lung disorders in calves and young cattle by changing the ventilation, usually by increasing ventilation and improving the circulation of fresh air.

Notably, the course of and the recovery from pulmonary disorders are in large part dependent on the climate. Suitable bedding, a well-ventilated environment and rest are important conditions for recovery.

Preventive measures:

- Provide bedding which is not too cold.
- Sufficient and proper ventilation: renew sufficient m^3/h with a proper air current pattern and without draught.
- Not too many animals per pen: animals/m^2 plus sufficient space: m^3/animal.
- No great disparity in age within one herd.

Reference

Andrea/Gelber: Süchtungskunde 1968-69.

DISCUSSION

W. Goldhorn *(CEC)* Where I come from in Germany there are hundreds of bull fattening units and they had all these problems. I support your opinion that there is a connection between ventilation, temperature and humidity. In winter when the temperature drops very low in Munich, a badly insulated stable will inhibit the ventilation rate because ventilation is linked with stable temperature. This means if the temperature drops greatly, then the ventilation will be very slow which will result in a very high relative humidity and encourage the presence of other toxic gases as well. As a result there is only one possible action, the ventilation must be switched on. However, then you get a very cold climate and draughts. The stables must be insulated much better to prevent this.

I would like to draw your attention to a point concerning straw bedding. This is not only a matter of regulating the temperature but has also something to do with feeling cosy and well. This improves the chances of fighting against all the bacteria and viruses. When first dealing with these problems, we advised the farmers to disinfect, and the best way to disinfect is to use slatted floors. However, we ran into more and more problems and began to draw the contrary conclusion and so advised giving more straw bedding. Even though this meant that disinfection was more difficult, the infection rate dropped significantly. We then added another possibility and I would add this to your preventive methods - a quarantine stable. It is not possible to stop calves that arrive from different farms bringing infectious agents with them. If you put these calves in a large stable with other adult bulls, the latter can easily pass on their infections to the calves especially since the calves arrive suffering great stress from the market, from transportation, from a change in nutrition, etc. Or, the contrary can happen, the calves can bring new viruses that can affect the adult herd. Heavy losses can be suffered in this manner. So, we introduced quarantine stables with straw bedding and this gave us the best results of all.

C. Holzhauer *(Netherlands)* I agree with you that young calves should not be brought into stables with older animals. They should be 4 - 5 months old before being put with older animals and also with other young calves. Good ventilation helps to remove pathogens and this also removes the need for such strong disinfection.

P.J. O'Connor *(Ireland)* At the moment we do not have great problems with IBR but we are keeping a very careful eye on it. I was not aware that such young calves were being affected.

C. Holzhauer We do not always see IBR in young calves. Sometimes in this farm there were cows which had had IBR but there were also cows that did not have it. So some of these young calves had no maternal immunity and yet in the same stable a cow would be brought that was calving, infected with IBR or with some symptoms of this disease. All the young calves which came into contact would become infected and would subsequently die. Sometimes as many as 12 or 14 calves have died. Often, however, IBR infection in young animals has no symptoms.

GENERAL DISCUSSION

P.J. O'Connor *(Ireland)* Dr. Bell made an interesting observation in his paper about an oesophageal groove and the closing of the same. He talked about teat size and about milk, the milk replacer spilling into the reticulum. Is there anything we could do about this? Is there anything that could be done about teat size, or is there a chemical that could be put into the feed replacer which would affect the oesophageal groove and so get the product right to the abomasum?

F.R. Bell *(UK)* There is some limited evidence that certain elements such as copper or sodium are effective in producing the reflex closure of the reticular groove. It looks as though it is the mouthing of the milk, the actual milk passing over receptors in the pharynx, that produces the closure. Certainly, there is very good evidence that if a calf drinks water from a bucket this will not close the groove. If the calf drinks milk from a bucket then it will close most of the time. If it drinks milk from a teat then it will close all, or nearly all, the time.

L.O. Fiems *(Belgium)* Dr. Blom, did you observe any difference in the occurrence of respiratory disease in group housed calves compared to individually housed animals?

J.Y. Blom *(Denmark)* All our boxes were numbered which was done in an attempt to learn the difference in incidence between those calves in group housing and those in individual pens. The calves were most at risk during the period when they were in the individual pens. Further analysis on this did not reveal anything of significance so we could not make any reasonable assumptions about the disease incidence in group pens compared to individual pens. You must remember that our calves were not from different sources but from the same herd which is a different situation compared to that of the fattening herd.

I would like to show you one more figure which deals with management problems. We talk a lot about the importance of the environment, and indeed this is important as Dr. Holzhauer has shown. The question of management is these days especially important since the herdsman has so much to do and is under so much economic pressure that he tends not to have enough time, or is perhaps not properly trained, to attend to his calves.

This figure shows the total calf mortality, that is the overall mortality from all herds, in relation to who attended the calves. This should emphasise the need to train.

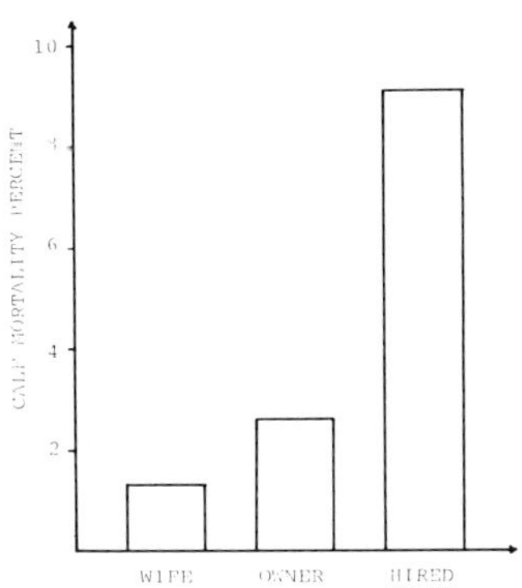

Fig. 1. Calf mortality as related to management.

H.C. Adler *(Denmark)* I will now conclude this session by bringing together a few points. There is no need in this session to pose the question "What has it to do with welfare?". We are in a situation where we have an animal industry which operates, in some countries, with a 15% mortality rate. There is also a high rate of morbidity, and this situation is caused by the way in which we have arranged this industry. This is not an acceptable situation, from a welfare or an economic viewpoint.

SESSION III

HOUSING OF CALVES

Chairmen: J.M. Bienfait and W. Sybesma

EFFECT OF FEEDING TECHNIQUES AND AGE AT WEANING ON THE PERFORMANCES OF BUCKET-FED AND SUCKLING REARED CALVES*

L.O. Fiems, Ch.V. Boucqué, B.G. Cottyn and F.X. Buysse

National Institute for Animal Nutrition
Scheldeweg 68, B-9231 Melle-Gontrode, Belgium.

ABSTRACT

Different rearing trials were carried out with newborn Belgian white-red male calves to investigate the impact of labour-saving systems, such as once daily feeding, an adapted feeding schedule for weekends, early weaning, automatic feeding and multiple suckling, on their performances.

No unfavourable effect was observed with once daily feeding and early weaning. These systems stimulated starter intake. Calves can be successfully weaned when they consume 0.5 kg concentrate per day. Omission of milk feeds during the weekend resulted in a significantly lower pre-weaning daily gain in one of the two trials, as did a poor rearing system with a combination of different labour saving techniques. The lower pre-weaning growth rate was largely compensated afterwards.

Automatic feeding did not improve daily gain and resulted in more problems with scouring when compared to bucket feeding. Multiple suckling calves showed a significantly higher liveweight gain during the rearing period. Afterwards the calves fed by pail revealed only a weak compensatory gain.

The early feeding of grass hay stimulated concentrate intake and consequently increased the growth rate.

INTRODUCTION

Some newborn calves are used for veal production. The remainder are kept as rearing calves. Rearing calves are not a goal in themselves, but they are of capital importance since heifer calves will later replace culled cows (Troccon and Journet, 1979) and bull calves are mostly used for beef production (Buysse and Boucqué, 1975).

Although heifer calves and bull calves are reared for different purposes, the management of both categories during their early life can be very similar. Since the forestomachs of the newborn calf are essentially non-functional the animals are dependent on colostrum during the first days of life, followed by a milk diet. Afterwards the development of the

*Communication of the Institute N° 470. CEC Seminar Calf Welfare, Brussels, 9-10 July, 1981.

ruminant forestomachs is influenced by the intake of concentrates and roughages (Warner et al., 1956).

Apart from the administered milk diet, concentrate and roughage, good health, optimal animal performance and acceptable feed cost are of major interest. Further, labour-saving feeding techniques are welcome as they meet the social needs of the cattle breeder.

EXPERIMENTAL PROCEDURE

Many rearing trials were carried out to study the following topics:

- once or twice daily feeding
- an adapted feeding schedule for weekends
- age at weaning
- weaning as a function of the calf starter intake
- automatic or bucket feeding
- a liberal versus a poor rearing system
- suckling or bucket feeding
- early hay administration

Male calves of the Belgian white-red dual-purpose breed were purchased in the market at an age of 8 to 10 days. In most experiments the rearing period lasted 112 days.

The impact of the feeding system on feed intake, daily gain, feed costs and calf health was investigated. It is difficult to find operational parameters to estimate animal welfare (van Putten, 1978).

A commercial milk replacer was always used. Water and grass hay were freely available, while the concentrate administration was mostly restricted to 3 kg/calf/day. Calf starter never contained any milk product. Concentrate and grass hay which had not been eaten were weighed once a week. The calves were housed in individual straw-bedded pens, except those fed with an automatic liquid feeder and the suckling calves.

Feed prices amounted to 30 BF/kg milk replacer, 3.5 BF/kg grass hay and 9.5 BF/kg concentrate.

RESULTS AND DISCUSSION

Once versus twice daily feeding

In a rearing trial involving 2 groups of 13 calves, the effect of once versus twice daily feeding of milk replacer was investigated. However, both groups received the same amounts of milk substitute. Weaning occurred after 8 weeks. The calf starter intake was restricted to 3 kg, while grass hay was freely available. The mean results are presented in Table 1.

TABLE 1

EFFECT OF ONCE OR TWICE DAILY FEEDING OF MILK SUBSTITUTE

Feeding schedule	Twice daily	Once daily	
Number of calves	13	13	
Initial weight (kg)	49.4	49.5	
Final weight (kg)	134.2	138.4	
Length of experiment (days)	112	112	
Daily gain (g)			
0 - 8 weeks	600	648	n.s.
0 - 16 weeks	757	794	n.s.
Feed intake (kg)			
Milk substitute	22.8	22.8	
Calf starter	190.7	197.1	
Grass hay	37.9	47.3	
Feed costs (BF/kg gain)	31.0	30.6	

n.s.: not significantly different ($P > 0.05$)

The daily gain was higher, although not significantly different, for the calves fed once daily, both during the milk period and during the total rearing period. There was no unfavourable effect on feed intake. On the contrary, one milk feed per day stimulated starter and hay consumption. This tendency had already started during the milk period, and is an explanation of the higher growth rate. Feed costs were not at all negatively affected with this labour-saving feeding technique

More information is cited by Eeckhout and Buysse (1971). A number of experiments have shown no detrimental effects due to daily feeding on the growth rate and the health of the calf (Owens and Stake, 1972; Reddy et al., 1971; Fieber, 1972; Hoden and Journet, 1972; Galton and Brakel, 1976). Once daily feeding required about 40% less labour in milk feeding than twice daily feeding (Galton and Brakel, 1976; Troccon and Giovanni, 1976).

An adapted feeding schedule for weekends

A calf rearer normally has to take care of the calves 7 days a week. The introduction of a free weekend can reduce this social problem. Two experiments were carried out. The calves were weaned after 8 or 5 weeks respectively. During the first two weeks the milk substitute was offered daily in two equal feeds at 0730 h and 1600 h. From the third week the experimental group received only one milk feed on Saturdays at 1100 h and no milk at all on Sundays. Half of the calves in both trials were individually penned, and the remainder housed in groups of four, always on a straw-bedded floor. Further detailed information has already been published by Boucqué et al. (1971).

The results in Table 2 indicated a variable effect of omitting three milk feeds per week. When the calves were weaned after 8 weeks the free weekend feeding schedule resulted in a higher but not significantly different daily gain, as well during the milk period as the whole rearing period. The lower intake of milk replacer was counterbalanced by a higher calf starter consumption for the experimental animals.

The second trial revealed a lower growth rate of the experimental group during the pre-weaning and the total rearing period. However, the difference was only statistically significant during the first five weeks. The higher milk substitute intake by the control animals was not compensated by a higher starter intake by the experimental group.

In both trials, feed costs were not negatively affected by applying the free weekend. Good results obtained with this kind of feeding system were also reported by Troccon and Giovanni (1976) and Troccon and Rossat (1978).

TABLE 2

EFFECT OF FEEDING MILK SUBSTITUTE 11 OR 14 TIMES PER WEEK

Weaning age	8 weeks		5 weeks	
Meals per week	14	11	14	11
Number of calves	22	23	31	34
Initial weight (kg)	48.1	48.2	47.6	47.2
Final weight (kg)	144.3	147.6	142.4	136.5
Length of experiment (days)	112	112	112	112
Daily gain (g)				
0 - weaning	$673^{a} \pm 39$	$700^{a} \pm 29$	$600^{a} \pm 23$	$506^{b} \pm 26$
0 - 16 weeks	$859^{a} \pm 26$	$888^{a} \pm 21$	$846^{a} \pm 24$	$797^{a} \pm 25$
Feed intake (kg)				
Milk substitute	31.0	25.5	21.3	18.0
Calf starter	203.9	213.1	213.8	206.9
Grass hay	33.8	32.2	22.8	20.1
Feed costs (BF/kg gain)	31.0	29.2	29.0	28.8

a, b: values on the same line bearing different superscripts are significantly different ($P < 0.05$).

Age at weaning

Eighty-four bull calves were divided into 6 groups and received a total amount of milk replacer varying between 23 and 28 kg. The milk diet was administered twice daily for 5, 6, 7, 8, 9 or 10 weeks. The calves had free access to concentrate and grass hay.

Although there were some statistical differences in pre-weaning daily gain, liveweight gain for the total rearing period was not adversely affected by early weaning. This means that the group with a lower growth rate during the milk period had a compensatory gain afterwards. Total feed intake was very similar for all groups, resulting in a similar feed cost. The most important results are summarised in Table 3. Further details have previously been published by Boucqué and Buysse (1967). When different amounts of milk products were fed we found a substitution of other feedstuffs for milk products, resulting in an equal liveweight gain (Boucqué and Buysse, 1966).

TABLE 3

EFFECT OF AGE AT WEANING ($\pm S_{\bar{x}}$)

	Age at weaning (weeks)					
	5	6	7	8	9	10
Number of calves	16	12	14	14	14	14
Initial weight (kg)	46.8	46.8	48.3	47.9	46.9	47.1
Final weight (kg)	136.5	139.4	139.9	139.8	131.6	139.6
Length of experiment (days)	112	112	112	112	112	112
Daily gain (g)						
0 - weaning	$545^{ac} \pm 40$	$671^{bc} \pm 59$	$527^{a} \pm 51$	$578^{ab} \pm 40$	$578^{ab} \pm 21$	$688^{b} \pm 43$
0 - 16 weeks	$801^{a} \pm 30$	$827^{a} \pm 53$	$818^{a} \pm 54$	$821^{a} \pm 34$	$756^{a} \pm 28$	$826^{a} \pm 45$
Feed intake (kg)						
Milk substitute	23.1	27.8	24.1	28.6	23.4	27.8
Calf starter	207.0	195.6	199.6	203.4	187.4	198.4
Grass hay	28.5	20.9	26.7	19.5	20.1	28.4
Feed costs (BF/kg gain)	30.8	29.9	29.6	31.1	30.1	30.5

a, b: values on the same line having different superscripts are significantly different ($P < 0.05$)

Weaning as a function of the calf starter intake

Since the forestomachs have to evolve from being non-functional in a newborn calf to being functional in older ruminants the following trial was conducted. Four groups of individually housed male calves were weaned either after eight weeks or when the weekly starter intake exceeded 5, 4 or 3 kg. This corresponded with a daily concentrate intake in the last week prior to weaning varying between 400 and 600 g, 600 and 800 g or more than 800 g. Calves which did not consume this amount were nevertheless weaned after 8 weeks. The milk replacer was administered twice daily for 4 weeks, and thereafter once daily. After the 4th week the 'free weekend' was applied. Grass hay was freely available, but the calf starter was restricted to 3 kg per day.

The results presented in Table 4 show that the calves weaned after 8 weeks had a higher daily gain, although the difference was not significant. The growth rate in general was rather low due to a relatively low concentrate intake (see the review - Table 9).

TABLE 4

WEANING AGE AS A FUNCTION OF CALF STARTER INTAKE

Weaning age	8 weeks	Weekly starter intake prior to weaning		
		>5 kg	>4 kg	>3 kg
Number of calves	10	10	11	11
Initial weight (kg)	47.3	45.7	48.8	46.5
Final weight (kg)	125.9	118.9	122.1	118.6
Weaning age (d)	56.0	39.2	31.1	29.3
Length of experiment (days)	112	112	112	112
Daily gain (g)$^{n.s.}$	702	653	654	644
Feed intake (kg)				
Milk substitute	22.1	17.1	13.6	13.0
Calf starter	169.8	177.2	187.0	183.2
Grass hay	31.4	18.9	19.8	21.5
Feed costs (BF/kg gain)	30.4	30.9	30.7	30.6

n.s.: not significantly different ($P > 0.05$)

Early weaning resulted in a lower intake of milk replacer but also in an increase of calf starter consumption. Once again this is in accordance with the results of Boucqué and Buysse (1966). More data about these experiments are mentioned by Eekhout and Buysse (1969).

Owen and Plum (1968) found no significant effect on weight gain when calves were weaned at 21 days of age or when the starter intake had reached 454 g per day during the last week prior to weaning. From experiments of Kertz et al. (1979) it became clear that an early consumption of calf starter is of major importance. Concentrate intake accounted for 65% of the variation in total liveweight gain during the first 28 days.

Automatic versus bucket feeding

Milk feeding can be reduced to the preparation of the milk replacer by automatic liquid feeding. In this trial the calves were fed by pail or by automatic feeder with unlimited access or with access limited to 2 h/day (reconstitution rate: 11%) and were weaned after 8 or 6 weeks. The calves fed automatically were group housed while those fed by pail were kept in individual pens.

Automatic feeding did not result in a higher daily gain compared with bucket feeding. In fact, the opposite was true. The same phenomenon was established with veal calves (Cottyn et al., 1969). The large variation of the liveweight gain in these groups is due to strong calves tending to obtain more milk. The growth rate was still diminished when daily access was reduced to 2 h. The highest intake of milk substitute was noted for calves fed *ad libitum* and this decreased when daily access was reduced. A high intake of milk substitute restrained the starter consumption.

Due to the abundant intake of milk substitute, 3 calves died of diarrhoea: one in each group, except in the group weaned after 8 weeks and with 2 h daily access. This negative aspect was also stated with veal calves (Cottyn et al., 1969). However, no detrimental health effects in calves fed *ad libitum* were mentioned by Lineweaver and Hafez (1969).

Furthermore, feed costs rose markedly with automatic liquid feeding. The most important results are summarised in Table 5 and more information is given by Cottyn and Boucqué (1972).

During recent years the *ad libitum* feeding of cold acidified milk substitute (Stobo and Roy, 1980) has been recommended as it is claimed that it is beneficial in reducing the incidence of scours.

Effect of a liberal versus a poor rearing system

The different feeding techniques of early weaning (after 5 weeks), once daily feeding and a free weekend were combined in one trial. The performances were compared with those of calves weaned after 10 weeks and fed twice daily without a free weekend. Except for a different amount and reconstitution rate of the milk substitute the administration of calf starter and grass hay was equal for both groups.

The results shown in Table 6 revealed a significantly lower daily gain during the rearing period for the poor feeding system. The lower consumption of milk substitute stimulated the concentrate intake, but this was not sufficient to equalise the liveweight gain.

Three calves in this group died, 2 of diarrhoea after 5 and 13 days respectively and one of bloat after 13 days. More scouring was also reported by Wood et al. (1971) when calves were fed once daily and with a weekly fast of 39 h. Wooden et al. (1968) mentioned reduced vigour, more pneumonia and slightly more scours in calves fed once daily.

The effect of the rearing system on growth capacity was investigated during the subsequent grazing and finishing period. Calves fed under the poor system during the rearing period had a higher daily gain, although this was not significantly different. The cumulative daily gain from birth to slaughter was 12 g lower, resulting in a liveweight difference at slaughter of 7.3 kg.

Effects of multiple suckling as against bucket feeding

The labour required in rearing calves can be reduced by suckling. A trial was therefore set up to compare the performances

TABLE 5

EFFECT OF AUTOMATIC MILK FEEDING VERSUS BUCKET FEEDING

Weaning age	8 weeks			6 weeks		
Feeding system Daily access (h)	Bucket	Automatic 24	2	Bucket	Automatic 24	2
Number of calves	14	10	10	12	11	9
Initial weight (kg)	47.9	48.6	45.2	46.8	47.9	45.8
Final weight (kg)	139.8	138.4	110.7	139.4	128.0	120.9
Length of experiment (Days)	112	112	112	112	112	112
Daily gain (g)	$821^{a} \pm 34$	$802^{a} \pm 57$	$585^{b} \pm 52$	$827^{a} \pm 53$	$715^{ab} \pm 56$	$671^{ab} \pm 65$
Feed intake (kg)						
Milk substitute	28.6	55.4	44.9	27.8	35.6	29.7
Calf starter	203.4	145.0	140.4	195.6	165.8	173.6
Grass hay	19.5	32.0	29.0	20.9	18.9	29.5
Feed costs (BF/kg gain)	31.1	35.1	42.5	29.9	33.8	35.2

a, b: values with different superscripts are significantly different ($P < 0.05$)

TABLE 6

EFFECT OF A LIBERAL VERSUS A POOR REARING SYSTEM ($\pm s_{\bar{x}}$)

Feeding schedule	Group I	Group II
weaning age	10 weeks	5 weeks
meals per week	14	6
Number of calves	35	32
Initial weight (kg)	46.9	47.1
Final weight (kg)	136.8	126.0
Length of experiment (days)	112	112
Daily gain (g)		
0 - 5 weeks	523 ± 21	400 ± 30**
0 - 10 weeks	763 ± 18	616 ± 22**
0 - 16 weeks	803 ± 14	705 ± 14**
Feed intake (kg)		
Milk substitute	38.8	15.0
Calf starter	182.9	200.1
Grass hay	23.0	21.0
Feed costs (BF/kg gain)	33.2	30.7
Daily gain (g)		
Pasture period	570	599 n.s.
Finishing period	1 308	1 318 n.s.
0 - slaughter	989	977 n.s.

**$P < 0.01$. n.s.: not significantly different ($P > 0.05$)

of multiple suckling calves with those of bucket fed calves. Milk substitute was administered for a period of 8 weeks, twice daily during the first two weeks and once daily with omission of the milk feed on Sunday after the third week. All calves had free access to grass hay and concentrate intake was restricted to 3 kg per day. The rearing period lasted about 20 weeks. More details have been reported by Boucqué et al. (1978b).

Suckling calves had a significantly higher growth rate ($P < 0.01$). Although labour-saving, the feed costs per kg gain of these calves were higher when the feed intake of the dams was taken into account. Gathering calves from different farms for multiple suckling can result in sanitary problems (Boucqué et al., 1978a).

After the rearing period the suckling calves and half the bucket fed calves were turned out to pasture. Both groups were given a supplement of dried sugarbeet pulp, but suckling calves could go on suckling. The other half of the bucket fed calves stayed indoors and received a concentrate diet to appetite. During the finishing period the same concentrate diet was fed to the three groups. Calves fed by pail realised only a weak compensatory gain after the rearing period. Daily gain from birth to slaughter amounted to 1 151 g for multiple suckling calves and was significantly higher compared to that of the bucket fed bulls whether on pasture or indoors. These results are presented in Table 7.

TABLE 7

EFFECT OF SUCKLING VERSUS BUCKET FEEDING

Feeding system	I	II	III
Rearing period	suckling	bucket	bucket
Pasture period	pasture	pasture	indoors
Finishing period	indoors	indoors	indoors
Number of calves (cows)	26 (8)	26	26
Initial weight (kg)	47.9	48.0	48.3
Final weight (kg)	168.7	140.3	142.5
Length of experiment (days)	136.3	134.5	134.1
Daily gain (g)	$885^{A} \pm 28$	$686^{B} \pm 23$	$702^{B} \pm 25$
Feed intake (kg)			
Milk substitute	-	24.2	23.7
Calf starter	131.1	228.0	225.6
Grass hay	-	40.2	43.6
Feed costs (BF/kg gain)	42.5	32.9	31.9
Daily gain (g)			
Pasture period	$1\,230^{A} \pm 31$	$856^{B} \pm 24$	$1\,234^{A} \pm 29$
Finishing period	$1\,321^{A} \pm 43$	$1\,390^{A} \pm 29$	$1\,091^{B} \pm 40$
0 - slaughter	$1\,151^{A} \pm 17$	$1\,004^{B} \pm 16$	$1\,040^{B} \pm 16$

A, B: means on the same line with different superscripts are significantly different ($P < 0.01$).

Effect of grass hay administration at different ages

There is a tendency to allow veal calves a small amount of straw. We studied the effect of roughage administration on calf rearing. Bull calves receiving the same milk diet and concentrate were fed grass hay from the first week or the ninth week respectively. Grass hay was of good quality (25% crude fibre) and was freely available for both groups.

Early feeding of grass hay resulted in a higher liveweight gain during the first eight weeks and the whole rearing period. This is due to the stimulated solid feed intake. These results are presented in Table 8.

TABLE 8

EFFECT OF GRASS HAY ADMINISTRATION AT DIFFERENT AGES

	Grass hay administration (weeks)	
	1 to 16	9 to 16
Number of calves	12	13
Initial weight (kg)	46.4	46.3
Final weight (kg)	142.9	135.5
Length of experiment (days)	112	112
Daily gain (g)		
0 - 8 weeks	629	517
9 - 16 weeks	1 094	1 077
0 - 16 weeks	862	797
Feed intake (kg)		
Milk substitute	22.8	22.7
Calf starter	225.7	203.6
Concentrate	31.4	15.8
Feed costs (BF/kg gain)	30.4	29.9

Several investigations emphasised the importance of roughage administration. It promoted saliva secretion by means of a greater eating and ruminating time, and resulted in a greater rumen buffering capacity. Furthermore, the total feed intake was also increased (Bakker, 1974; Leibholz and Russell, 1980; Thomas et al., 1980). The greater buffering capacity may be responsible for a diminished rumen acidity. Hyperacidity had frequently been implicated as contributing to rumen parakeratosis, ulceration of the rumen wall and liver abscesses (Breukink, 1981). Any one of these conditions may depress intake and growth rate. Maize silage can be successfully substituted for grass hay. (Fiems et al., 1979).

CONCLUSIONS

Once daily feeding, omission of milk feeds on Sunday, early weaning, automatic liquid feeding and multiple suckling can be advocated as labour-saving feeding techniques. Nevertheless, feed costs are increased by automatic feeding and suckling. The main results are recapitulated in Table 9.

In our trials we found no detrimental effects of once daily feeding and early weaning on daily gain and calf health. However, weak calves can be weaned better when daily starter intake reaches about half a kg.

Applying a free weekend resulted in a variable effect on pre-weaning daily gain, but liveweight gain for the whole period was not significantly different. Once daily feeding and a free weekend may be postponed in sick calves.

Automatic liquid feeding did not improve daily gain and resulted in digestive upsets.

The combination of different labour-saving feeding techniques resulted in a significantly lower growth rate. However, the lower liveweight was largely compensated after the rearing period. This rearing system can only be applied with caution in healthy calves.

Bucket fed animals showed a significantly lower daily gain compared to multiple suckling calves during the rearing period, and only achieved a small compensatory gain. Purchasing calves from different farms for multiple suckling can result in sanitary problems.

Early hay administration stimulates total solid feed intake and daily gain.

ACKNOWLEDGEMENTS

The authors are indebted to Dr. R. Moermans of the Biometric Unit of the Centre of Agricultural Research at Merelbeke for

TABLE 9

SUMMARY OF THE MAIN REARING RESULTS

Rearing system	n	Liveweight interval (kg)	Daily gain (g)	Feed intake (kg)			Feed costs (BF/kg gain)
				Milk subst.	Starter	Grass hay	
Once daily feeding	13	49.5 - 138.4	648	22.8	197.1	47.3	30.6
Twice daily feeding	13	49.4 - 134.2	600	22.8	190.7	37.9	31.0
Weaned at 8 weeks							
14 meals per week	22	48.1 - 144.3	859	31.0	203.9	33.8	31.0
11 meals per week	23	48.2 - 147.6	888	25.5	213.1	32.2	29.2
Weaned at 5 weeks							
14 meals per week	31	47.6 - 142.4	846	21.3	213.8	22.8	29.0
11 meals per week	34	47.2 - 136.5	797	18.0	206.9	20.1	28.8
Weaning age: 10 weeks	14	47.1 - 139.6	826	27.8	198.4	28.4	30.5
9 weeks	14	46.9 - 131.6	756	23.4	187.4	20.1	30.1
8 weeks	14	47.9 - 139.8	821	28.6	203.4	19.5	31.1
7 weeks	14	48.3 - 139.9	818	24.1	199.6	26.7	29.6
6 weeks	12	46.8 - 139.4	827	27.8	195.6	20.9	29.9
5 weeks	16	46.8 - 136.5	801	23.1	207.0	28.5	30.8
Weaning after 8 weeks	10	47.3 - 125.9	702	22.1	169.8	31.4	30.4
Weaning after 5 kg starter/week	10	45.7 - 118.9	653	17.1	177.2	18.9	30.9
Weaning after 4 kg starter/week	11	48.8 - 122.1	654	13.6	187.0	19.8	30.7
Weaning after 3 kg starter/week	11	46.5 - 118.6	644	13.0	183.2	21.5	30.6

TABLE 9 (cont.)

Rearing system	n	Liveweight interval (kg)	Daily gain (g)	Feed intake (kg) Milk subst.	Starter	Grass hay	Feed costs (BF/kg gain)
Weaning after 8 weeks							
Bucket feeding	14	47.9 - 139.8	821	28.6	203.4	19.5	31.1
Automatic: 24/24 h	10	48.6 - 138.4	802	55.4	145.0	32.0	35.1
Automatic: 2/24 h	10	45.2 - 110.7	585	44.9	140.4	29.0	42.5
Weaning after 6 weeks							
Bucket feeding	12	46.8 - 139.4	827	27.8	195.6	20.9	29.9
Automatic: 24/24 h	11	47.9 - 128.0	715	35.6	165.8	18.9	33.8
Automatic: 2/24 h	9	45.8 - 120.9	671	29.7	173.6	29.5	35.2
Liberal rearing system	35	46.9 - 136.8	803	38.8	182.9	23.0	33.2
Poor rearing system	32	47.1 - 126.0	705	15.0	200.1	21.0	30.7
Multiple suckling	26	47.9 - 168.7	885		131.1		
Bucket feeding	26	48.0 - 140.3	686	24.2	228.0	40.2	32.9
Bucket feeding	26	48.3 - 142.5	702	23.7	225.6	43.6	31.9
Hay from 1st week	12	46.4 - 142.9	862	22.8	225.7	31.4	30.4
Hay from 9th week	13	46.3 - 135.5	797	22.8	203.6	15.8	29.9

the statistical analysis. They also gratefully thank Mr. A. Hertegonne and Mrs. M. de Bruycker for their skilled technical assistance.

REFERENCES

Bakker, Y.Tj., 1974. De groei, voederopname en gezondheid van fokkalveren bij uitstel van de ruwvoederverstrekking en bij toepassing van verschillende ruwvoeders, krachtvoeders en krachtvoederniveaus. De Schothorst, Proefverslag N° 54.

Boucqué, Ch.V. and Buysse, F., 1966. L'utilisation de différentes quantités de lait artificiel dans les régimes d'élevage de veaux femelles et l'influence sur l'évolution ultérieure du poids de ces veaux. Revue Agric., Brussels 19, 421.

Boucqué, Ch.V. and Buysse, F., 1967. Essai de sevrage précoce de veaux males destinés à la production intensive de viande. Revue Agric., Brussels 20, 415.

Boucqué, Ch.V., Buysse, F.X. and Cottyn, B.G., 1977. The effect of giving 11 or 14 feeds of milk substitute per week to early weaned calves. Anim. Prod. 13, 613.

Boucqué, Ch.V., Fiems, L.O. and Buysse, F.X., 1978a. Plusieurs veaux par vache allaitante de race pure à double fin. Revue Agric., Brussels 31, 21.

Boucqué, Ch.V., Fiems, L.O. and Buysse, F.X., 1978b. Influence de l'aliment naturelle ou au moyen d'un aliment d'allaitement chez les veaux d'élevage, et du système alimentaire après l'élevage sur les performances des taurillons à viande. Revue Agric., Brux. 31, 225.

Breukink, H.J., 1981. De ontsporing van de pensfermentatie. Tijdschr. Diergeneesk. 106, 185.

Buysse, F.X. and Boucqué, Ch.V., 1976. Present and future beef production systems in Belgium. In: Improving the nutritional efficiency of beef production. Com. Eur. Com., Eur. 5488e, p. 15.

Cottyn, B.G., Boucqué, Ch.V. and Buysse, F.X., 1969. Expériences sur le plan d'allaitement automatique des veaux de boucherie. Revue Agric., Brussels 22, 57.

Cottyn, B.G. and Boucqué, Ch.V., 1972. Expériences sur le plan d'allaitement automatique des veaux d'élevage. Revue Agric., Brussels 25, 807.

Eekhout, W. and Buysse, F., 1969. Le sevrage de veaux mâles en fonction de l'ingestion de concentrés. Revue Agric., Brux. 22, 1603.

Eeckhout, W. and Buysse, F., 1971. L'influence de la distribution de lait reconstitué, en un ou en deux repas pas jour, à des veaux mâles destinés à la production intensive de viande. (Troisième communication). Revue Agric., Brussels 24, 1351.

Fieber, M.W., 1972. Effect of once-a-day versus twice-a-day feeding of whole milk and milk replacer on young Holstein calves. J. Dairy Sci. 55, 703 (Abstr.).

Fiems, L.O., Boucqué, Ch.V., Cottyn, B.G. and Buysse, F.X., 1979. L'ensilage de mais en tant que fourrage grossier unique pour veaux d'élevage. Revue Agric., Brussels 32, 1139.

Galton, D.M. and Brakel, W.J., 1976. Influence of feeding milk replacer once versus twice daily on growth, organ measurements, and mineral content of tissues. J. Dairy Sci. 59, 944.

Hoden, A. and Journet, M., 1972. Distribution à des veaux d'élevage de lait reconstitué en 1 ou 2 fois par jour de la naissance au sevrage. Bull. Tech. CRZV 7, 23.

Kertz, A.F., Prewitt, L.R. and Everett, J.P., 1979. An early weaning calf program: summarization and review. J. Dairy Sci. 62, 1835.
Leibholz, J. and Russell, C.L., 1978. Chaffed and ground straw and lucerne in the diet of the early weaned calf. Anim. Prod. 27, 171.
Lineweaver, J.A. and Hafez, E.S.E., 1969. Feed intake and performance in calves fed *ad libitum* and four times daily. J. Dairy Sci. 52, 2001.
Owen, F.G. and Plum, M., 1968. Performances of calves as influenced by three criteria for weaning: age, weight gain or feed intake. J. Dairy Sci., 51, 971 (Abstr.).
Owens, M.J. and Stake, P.E., 1971. Once versus twice daily milk feeding of dairy calves. J. Dairy Sci., 54, 801 (Abstr.).
Reddy, C.S., Stiles, D.A., Olson, H.H. and Lee, D.D., 1971. Evaluation of milk replacer feeding frequency and protein source in a calf starter. J. Dairy Sci., 54, 801 (Abstr.).
Stobo, I.J.F. and Roy, J.H.B., 1980. Effect of acidified milk substitutes on health and growth of the calf. Anim. Prod., 30, 457 (Abstr.).
Thomas, D.B., Hinks, C.E. and Gilchrist-Shirlaw, D.W., 1980. The importance of roughages in relation to performances, feed intake and rumen fermentation patterns in early weaned calves. Anim. Prod., 30, 468 (Abstr.).
Troccon, J.L. and Giovanni, R., 1976. Suppression de l'allaitement du dimanche aux veaux d'élevage. Bull. Techn. CRZV, 25, 51
Troccon, J.L. and Rossat, P., 1978. Distribution de 6 repas lactés par semaine à des veaux pie-rouge. Bull. Techn. CRZV, 34, 23.
Troccon, J.L. and Journet, M., 1979. Vitesse de croissance optimale et alimentation des génisses. Bull. Techn. CRZV, 37, 5.
van Putten, G., 1978. Referentiesystemen in de teogepaste ethologie. Landbouwk. Tijdschr. 90, 118.
Warner, R.G., Flatt, W.P. and Loosli, J.K., 1956. Dietary factors influencing the development of the ruminant stomach. J. Agr. Food Chem., 4, 788.
Wood, A.S., Barley, H.S. and MacLeod, G.K., 1971. Evaluation of imposing a weekly fast on calves receiving a milk replacer diet once and twice daily: protein and energy utilization. J. Dairy Sci., 54, 405.
Wooden, K.G., Speicher, J.A. and Huber, J.T., 1968. Effect of feeding systems on feed and labor costs and rates of gain in dairy calves. J. Dairy Sci., 51, 971 (Abstr.).

DISCUSSION

W. Goldhorn *(CEC)* What is a 'free weekend'?

L.O. Fiems *(Belgium)* In this trial 'free weekend' meant that three meals were omitted. Normally the calves were fed twice a day, at 7.30 am and at 4.00 pm. On Sundays, the experimental group received only one milk feed at 11.00 am and no meals. There were only 11 milk feeds per week, but they received a normal amount of hay and concentrate.

N. Steenkamer *(Netherlands)* During these weekends, did you go to the units to hear and listen to the calves? We did the same trials and I had to stop them after a few weeks because I could not stand it. Perhaps I am oversensitive, but I could not do this combination of a weekend off-feeding plus once-a-day feeding, six milk feeds a day. Labour-saving is fine but I would say that this goes too far. When talking about welfare, we mention single pens, etc.; let us also look at the effects on welfare of this type of labour-saving system.

<u>L.O. Fiems</u> Yes, I agree, the animals did cry a great deal but I think the same phenomenon is observed when animals are weaned. During the first few days after weaning all the animals cry but this stimulates the intake of concentrates, and after a few days the crying ceases. We have examined different systems - gradual weaning and abrupt weaning. When calves were weaned after eight weeks, the first pattern was that there was an increasing amount of milk substitute, then the calves were abruptly weaned. Another pattern was that the calves received the same amount during the whole milk diet period. In the final trial there was an increase and after half the period there was a decrease. However, with all these feeding patterns we observed the same phenomena and so could not conclude that one or another of these patterns was better or worse.

THE EFFECT OF REARING SYSTEMS ON THE DEVELOPMENT OF BEHAVIOUR IN CALVES

A.J.F. Webster and Claire Saville

Department of Animal Husbandry,
University of Bristol, UK.

INTRODUCTION

It is a safe assumption, provided that one does not extend the argument to man, that the development of behaviour in an animal is determined both by its genetics and by its environment. One would therefore expect the development of behaviour in a young calf to be determined by its breed, its sex and the system of management imposed upon it in the first weeks of life. A systematic study has been made of the development of calf behaviour in five different systems which reflect those systems seen in commercial practice in the UK. The rearing systems included in this study were as follows:

1. Suckler cows with their calves at pasture.
2. Early weaned calves, penned individually and bucket-fed to weaning, thereafter reared in groups.
3. Early weaned calves, group penned and teat-fed cold, acidified milk replacer to weaner.
4. Veal calves, bucket-fed in individual wooden crates without bedding.
5. Veal calves, teat-fed warm milk from machines and group penned in straw yards.

Observations were made on calves at 2, 6, 10 and 14 weeks of age. The main purpose of this experiment was to evaluate the behaviour of veal calves in the context of more conventional rearing systems. The greatest number of observations available for analysis relate therefore to systems 4 and 5. Our observations on systems 1 - 3 have been made, in each case, on only two farms. The analysis of this study is incomplete and so this paper can only be described as a report on work in progress. Moreover, it is clear that within each of these systems there are likely to be variations in the nature and quality of stockmanship which would affect the behaviour of the animals. A comprehensive description of the effect of husbandry systems on the development of behaviour in calves requires a far larger experiment than we have been able to achieve so far. Fortunately, such an investigation, based upon the knowledge acquired in our preliminary study, is now being undertaken by the Veterinary Division of the UK Ministry of Agriculture, Fisheries and Food.

METHODS

The evolution of behaviour within each system has been deduced from observations of spontaneous behaviour made over 4 h during the period of daylight. These observations were timed to include both quiet and busy periods on the farm and one feeding period where appropriate. These observations have been augmented by a smaller set of continuous 24 h records obtained using teams of trained observers and by the use of 'stop-frame' cinephotography. In the first series of experiments each observer was presented with forms listing 23 spontaneous behaviour patterns. He or she then scanned the entire group of calves at 5 min intervals and recorded the number of calves engaged in a particular activity at the instant that that activity was being scanned. Such observations are then simply re-expressed in terms of the percentage time spent by calves on each activity and statistical analyses have been made on 30 min averages. In recent experiments designed to examine specific aspects of behaviour in more detail, calves have been scanned every 2 min.

An experiment was also designed to quantify the response of calves to a variety of stimuli arising from contact with man. Six stimuli were applied at random (slow approach, slight movement, sudden movement, whistle, shout and touch) and the response of individuals within groups of calves was scored on a points basis ranging from +2 for active interest through 0 for indifference to -3 for extreme fear.

RESULTS AND DISCUSSION

The ethogram

Table 1 summarises the major activities for calves in each of the rearing systems at 10 weeks of age. This table is intended

TABLE 1

SPONTANEOUS BEHAVIOUR OF CALVES IN COMMERCIAL UNITS AT 10 WEEKS OF AGE (PERCENTAGE TIME IN VARIOUS ACTIVITIES DURING 4 h OF DAYLIGHT)

	Systems				
	Suckled calves	Early weaned		Veal calves	
		buckets	teats	buckets	teats
Standing	27.0	13.5	14.8	20.4	15.4
Lying	13.1	37.3	32.0	40.6	54.4
'Sleeping'	2.8	9.1	5.8	3.2	15.9
Eating	25.2	32.6	34.7	None	4.4
Drinking	4.3	1.4	1.8	2.1	6.4
Ruminating	4.8	13.3	14.1	0.9	8.2
Chewing pen etc.	6.0	2.0	2.8	21.4	7.9
Grooming	2.4	5.7	6.1	11.0	4.9
Locomotion	7.4	3.3	10.5	3.2	4.2

only as an illustration of a complete set of results which will be analysed and published elsewhere. Nevertheless it immediately suggests profound differences between groups of calves in oral behaviour, resting behaviour and activities grouped under the general heading of locomotion. A more detailed analysis of these facets of behaviour based upon these observations in commercial units and 24 h observations made in the experimental calf rearing unit at the University of Bristol appears below.

Oral behaviour

Table 2 summarises 4 h sets of observations of oral behaviour in calves in commercial units at 10 weeks of age. In this case the two early weaned groups have been pooled because at this stage they were both being managed similarly in follow-on pens. Not surprisingly, the early weaned calves spent most time eating solid food. They also spent far longer ruminating than any other group. Veal calves with access to straw in yards ruminated almost twice as long as suckled calves at pasture. Veal calves in crates without access to straw ruminated hardly at all and that which was observed can best be described as pseudo-rumination. The high value for rumination time in the

TABLE 2

ORAL BEHAVIOUR OF CALVES AT 10 WEEKS OF AGE (PERCENTAGE TIME IN ACTIVITIES, 4 h OBSERVATIONS DURING DAYLIGHT)

	Suckled calves	Early weaned	Veal calves	
			straw yards	crates
Eating	25.2	33.6	4.4	None
Drinking	4.3	1.6	6.4	2.1
Ruminating	4.8	13.7	8.2	0.9
Grooming self	2.0	5.2	3.6	10.7
Sucking other calves	None	0.3	0.6	0.3
Licking or chewing pen etc.	6.0	2.2	7.3	21.4
Total	42.3	56.6	30.5	35.4

early weaned calves can be attributed to the precocious removal of milk or milk replacer from these animals; in this their oral behaviour can be said to be 'less normal' than that of veal calves in straw yards. Veal calves in crates spent much longer than calves in any of the other groups in grooming such parts of their body as they could reach with their mouths and tongues. The incidence of sucking the navels, prepuce or teats of other calves was very low for all groups. The highest incidence of 'vice' of this nature was observed in bucket-fed calves at 6 weeks of age (1.6%), i.e. immediately after weaning. Most of the calves that we observed spent quite a lot of time licking or chewing inanimate and apparently indigestible objects. Suckled calves at pasture spent about 6% of the time licking or chewing fence posts or wood in the hedgerows. Veal calves in straw yards spent an almost identical time licking the surrounds of their enclosure. Veal calves in crates which could neither eat nor ruminate spent abnormally long periods grooming themselves or licking or chewing their wooden pens. However, the total proportion of their time spent in oral activity was not abnormally long although, by our criteria, much of that activity would appear to have been purposeless.

Table 3 summarises observations made over 24 h on feeding behaviour of calves in straw yards with nominally free access to milk replacer from a teat dispenser. Each group of 15 calves had access to two teats. The groups of Friesian bulls

TABLE 3

FEEDING BEHAVIOUR OF MACHINE-FED CALVES AT 8 WEEKS OF AGE IN STRAW YARDS (15 CALVES AND TWO TEATS PER GROUP)

	Friesian bulls	Hereford x Friesian heifers
Body weight (kg)	96	76
No. of drinking sessions/24 h	8.1 (± 2.2)	6.8 (± 1.7)
Mean duration of drinking session (min)	7.1 (± 1.6)	7.1 (± 1.6)
Total drinking time (min/24 h)	57.0 (± 13)	48.0 (± 10)
Time 'fighting' at teat (min/24 h)	5.0 (± 3.0)	19.8 (± 9.0)
Liquid consumption per session (l)	1.91	2.33
Liquid/session/kg body weight (%)	1.98	3.07

and Hereford x Friesian heifers reflect the types of calves most commonly reared for veal in the UK. On average calves drank 7 - 8 times a day for about 7 min. The Hereford x Friesian heifers drank a greater proportion of their body weight per session and were far less 'polite', spending four times as long fighting for access to a teat. However there was no evidence to suggest that any individuals were failing to drink to appetite. It is, however, undeniable that veal calves drinking *ad libitum* in straw yards tend to show more variable weight gains than when they are individually confined and rationed to two bucket feeds per day. It has also proved very difficult to rear veal calves from straw yards to as high weights as can be achieved in crates. It appears that, in the latter stages of growth, veal calves bucket-fed twice a day are prepared to drink more milk than those animals with free access to teats (Table 4). This was particularly noticeable in the case of veal calves in crates but with free access to individual teats. These observations tentatively suggest that the appetite of the bucket-fed calf is abnormally high because the milk is consumed so rapidly that normal satiety mechanisms cannot be invoked. The question is not merely academic since we have had to sell a number of teat-fed calves 'unfinished,' i.e. not fat enough to receive quality payments. This interaction between feeding behaviour and profitability remains one of the most important unsolved problems of the straw yard system.

TABLE 4

CONSUMPTION OF MILK POWDER BY VEAL CALVES (kg/100 kg BODYWEIGHT)

	Age in weeks			
	2-6	6-10	10-14	14-16
Crated calves (Friesian bulls)				
Bucket fed	1.76	2.30	2.10	1.72
Nipple fed	1.71	2.20	1.93	1.28
Machine-fed calves in straw yards				
Friesian bulls	2.36	2.43	1.80	-
Hereford x Friesian heifers	2.68	2.51	2.01	-

Resting behaviour

Table 1 and Table 5 which are based on 4 h of measurement show that, at least during the hours of daylight, suckled calves at pasture spent much less time lying down than calves in any of the other groups. Table 5, which summarises observations at 10 - 14 weeks of age, shows that veal calves in straw yards lie down longer than those in crates. There was in fact a marked difference in lying time between 10 and 14 weeks of age for veal calves in crates. At the latter age and greater weight they appeared to be not only uncomfortable while lying but experienced some difficulty in changing position. The crates for these veal calves were 70 cm wide; most European crates are between 62 - 70 cm wide. It was, of course, impossible for veal calves in crates to lie on their side. Veal calves in yards lie on their side for 5% of the time. However, early weaned calves at this age lie on their sides for only 0.5% of the time and suckled calves not at all (at least during daylight). Table 5 suggests that denying the calf the right to lie on its side is not a severe insult to normal behaviour patterns. However, veal calves in crates over about 12 weeks of age were unable to adopt the classic sleeping posture with their heads tucked into their sides. A more valid criticism of the veal calf crate is therefore that it denies the larger calf the opportunity to sleep in comfort. Unless calves are reared to greater weights than at present this could be achieved by a crate width of 80 cm.

TABLE 5

SOME ASPECTS OF BEHAVIOUR IN CALVES AT 10 - 14 WEEKS OF AGE (AVERAGE PERCENTAGE TIME SPENT IN THE FOLLOWING ACTIVITIES; OBSERVATIONS DURING 4 h OF DAYLIGHT)

		Veal calves	
	Suckled	straw yards	crates
Lying, on brisket	13.1	46.0	33.6
on side	none	5.0	impossible
'Sleeping'	2.8	15.1	2.6
Run, play or fight	0.1	4.2	0.8
Cross grooming	0.4	1.0	2.0
Mounting	none	0.6	impossible

These observations on resting behaviour must be treated with some caution. Not only do they not cover the entire 24 h period but one must assume that the presence of the observer affected the behaviour of the calves to a degree that probably differed between systems. Table 6 summarises a range of behaviour patterns of calves based on 24 h observations. In the case of veal calves these were obtained by cinephotography in such a way as not to disturb the animals. When conditions were quiet and the animals were undisturbed, veal calves (at 10 weeks of age) in crates lie down longer than any of the other groups and their sleeping time also appeared to be normal. Table 6 should also be compared with Table 2, in the context of rumination behaviour. Suckled calves ruminated mostly at night. Veal calves in crates tended to do most of their compulsive licking or chewing of their crates in the daytime (or when they were disturbed). Veal calves in crates given 200 g/day chopped straw ruminated normally. Since the provision of some chopped straw also prevents some of the pathological abnormalities of rumen and abomasal development in veal calves and has no deleterious effect on food conversion efficiency (Webster and Saville, 1981), there can be no justification for denying veal calves the right to some solid food.

TABLE 6

SOME ASPECTS OF BEHAVIOUR OF 10 WEEK OLD VEAL CALVES AND SUCKLED CALVES AT PASTURE (PERCENTAGE TIME IN VARIOUS ACTIVITIES, BASED ON 24 h OBSERVATIONS)

	Suckled calves	Veal calves		
		straw yards	crates	
			no straw	200 g straw
Lying, on brisket	46.3	60.1	67.1	
on side	0.1	5.2	impossible	
'Sleeping'	19.5	21.3	16.4	
Locomotion	4.7	3.3	3.4	
Play or fight	0.1	2.1	impossible	
Cross grooming	0.3	1.1	2.5	
Mounting	none	0.2	impossible	
Ruminating	12.4	11.6	1.2	9.7
Grooming self	1.4	6.4	5.6	3.4
Licking or chewing pen etc.	2.0	3.5	13.8	10.6

Social behaviour

Suckled calves at pasture spent very little time indulging in social behaviour with other calves compared with veal calves in straw yards. In the Bristol unit veal calves in crates at 10 weeks of age and over spent 2% of the time grooming each other around the front over the top of adjacent pens. Implantation of bull calves or heifers with anabolic steroids (Implixa BM or BF respectively, Hoechst, UK) doubled the amount of time that calves in yards spent running, fighting, playing or mounting one another. This increase in activity appeared to reach a peak in the first week after implantation and returned thereafter to normal over a further period of about 3 weeks.

Response to man

The results of the experiment designed to evaluate the response of calves to the presence of man are summarised very briefly in Table 7 which presents overall scores for groups of calves at two ages. In its most simple form one can interpret

TABLE 7

OVERALL RESPONSE SCORE OF CALVES TO EXTERNAL STIMULI. CALVES ARE SCORED IN THEIR RESPONSE TO A NUMBER OF STIMULI VARYING FROM QUIET APPROACH TO SUDDEN NOISE AND MOVEMENT; THE MORE NEGATIVE THE OVERALL SCORE THE MORE TIMID THE RESPONSE.

	Age (weeks)	
System	2-4	10-14
Suckled calves at pasture	-54	-42
Early weaned calves, buckets to weaning	-12	-35
teats to weaning	-50	-34
Veal calves, crates	-48	-86
straw yards	-24	-14

the results of this experiment by saying that the more negative the overall response the more timid the animal. Table 7 shows that suckled calves at pasture generally reacted timidly to these stimuli in a way that was not influenced by age. Early weaned calves, bucket-fed, were tame while on bucket but became more timid after being turned out into follow-on pens. Veal calves in straw yards tended to become tamer with time. Veal calves in crates were similar to other groups under 4 weeks of age but showed increasingly fearful responses as they grew older. There is little point in speculating too elaborately on the reasons for this but it is relevant to mention that such calves are almost permanently out of contact with the sights and sounds of farm activities, the extreme restriction imposed upon their movement gives them no opportunity for quiet, evasive action and the noise generated by one animal reacting violently in a wooden crate tends to spread the panic.

CONCLUSIONS

It would be a mistake to attempt too many definitive conclusions at this stage. This is partly because the analysis is incomplete. For example, we do not yet know how accurately one can predict 24 h behaviour from observations made during 4 h of daylight. It is also highly probable that the behaviour of calves within any one system will be markedly affected by the quality of stockmanship. Nevertheless we can draw three firm conclusions which have a bearing on the keeping of veal calves in crates, which is the only system that has received severe criticism on the grounds of behavioural deprivation.

1. Although veal calves in crates spend an abnormally long time in apparently purposeless oral behaviour (chewing crates, etc.) the total time spent in oral activity is not excessive. However, since the provision of a small amount of forage stimulates normal rumination, reduces the risk of enteric disorders and does not reduce the efficiency of growth, there is a strong case for legislation designed to ensure that all calves over 2 weeks of age are not denied access to solid food.
2. Veal calves in crates become increasingly and abnormally apprehensive in response to stimuli normally associated with the presence of man.
3. The narrow veal calf crate obviously denies the calf the opportunity to lie on its side but this does not appear to constitute a severe behavioural deprivation. The conventional veal crate does, however, present the larger calf from adopting a normal sleeping position. At the very least, veal calf crates should have a minimum width of 80 cm. However, there are no health advantages to housing calves over 4 weeks of age in individual crates (Webster and Saville, 1981) so there can be no objections on grounds of health or welfare to a ban on total isolation of calves from birth to slaughter.

ACKNOWLEDGEMENT

Claire Saville was supported by a grant from the Farm Animal Care Trust.

REFERENCE

Webster, A.J.F. and Saville, Claire, 1981. Rearing of veal calves. Proc. UFAW Symposium, 'Alternatives to intensive husbandry'. Universities Federation for Animal Welfare, Potters Bar.

DISCUSSION

V. Jensen *(Denmark)* Why did you not use video in your experiments? With video you can get much better 24 h observations with the animals undisturbed by man.

A.J. Webster *(UK)* We could only use the equipment that was already available for financial reasons. Our observational techniques did not disturb the calves. The presence of man undoubtedly disturbs some groups of calves more than others, but man's presence does not disturb the veal calves kept in straw yards. The presence of the observer disturbs the veal calves in the crates for two reasons, partly because they are unused to man and partly because you have to move around to see all the calves. That machine runs up and down the line of crates, sees four crates and moves somewhere else.

A.A. Jongebreur *(Netherlands)* What was meant by 'lying on the side'? Have you any explanation for the difference between the results for the time found for lying on the side in suckling calves on the pasture and the veal calves in the straw yard?

A.J. Webster The main reason for that is thermo-regulation. Our observations of pasture were done mostly during the autumn and winter in England, so they were lying on the brisket with their feet tucked up to keep warm. Now I have done many measurements on the heat production of veal calves and I know that nothing produces more heat - at 10 weeks of age a veal calf is producing 600 watts. So a veal calf in a straw yard is a very hot animal indeed and the reason they lie on their sides is to dissipate as much heat as possible. Further analysis of this shows that the proportion of calves lying on their sides is a function of air temperature and age, because the bigger the calf the hotter it is. I am sure that this is the reason. This also supports Dr. van Putten's observations.

N. Steenkamer *(Netherlands)* Making a comparison between four different systems looks well on paper but how scientific is this?

A.J. Webster May I go back to what I said at the beginning. This is a description of methodology, if you like, for the study of the effect of a rearing system on the behaviour of calves. I am presenting the consequences of the methodology. We would never claim more than that this is what happened to our calves. We were sufficiently convinced that the methodology was all right, as was the Ministry of Agriculture, that they have now taken it up with 72 collaborating farms with either Friesians or Hereford-Friesians. The farms were carefully, not randomly, selected within each system so that the husbandry system within each group is as close as possible. Next year or so we will produce a document on the behaviour of calves in the south-west of Britain using this methodology but trying to answer these very real criticisms of yours.

N. Steenkamer You offered straw to calves in single pens and you were talking about a consumption of 200 g/day. From what age did they consume this 200 g and what type of straw was it? There are enormous variations, but our experience is that with barley straw we have to wait until the tenth week for them to consume 100 g. The straw must be very palatable for the calves to consume 200 g/day.

A.J. Webster It was spring barley straw in those trials, but not all of them ate 200 g. They were offered 200 g, but some of them never ate 200 g. We are now giving them, mostly on Dutch advice, a coarse straw-based pellet that we have made ourselves - straw, starch and whey protein. They all eat this within about 2 h at 6 weeks of age. They enjoy this and also it is not alkaline treated straw.

J.P. Signoret *(France)* I was most interested in the fact that you have attempted to quantify the reactions of the animal towards man. My question is mainly directed to other people here: are there similar observations made in other experiments concerning the effect of housing and management conditions on the reactions of the animal towards man? This is not only an obvious welfare criterion of the animal but could also be very important when transporting the animals, for instance. It could be an important trend both for practice and welfare.

H. Eichinger *(FRG)* When we do our treadmill exercises every two or three years with 100 or more calves, we have noticed that it is much easier to exercise a milk-replacer fed calf than one kept in a pen. This is probably because they are fed and handled by man and so are adapted to man. The caged calf has almost no opportunity to move and walk in a certain way but they walk extremely well on the treadmill while others often refuse and have a longer learning period than the singly caged animal. I think this is because their contact with humans is more intensive.

A.J. Webster I must stress again that those were observations on commercial units. If we did this to our own calves we would get a completely different pattern because all our calves are tame, even the ones in boxes since they see us all the time. You must make these observations on units where the calves are not expecting you to arrive.

GENERAL DISCUSSION

W. Sybesma *(Netherlands)* We have just had two papers presented on the rearing system. However, the two approaches were different and I think this shows the different possibilities for judging certain production systems. Professor Webster told us that methodology is very important in judging a system in practice. I think we can all agree that when we have a certain system, we should check it in practice. The problem is that conclusions must be drawn although not for some years to come. I should like to hear from Roy Moss about the policy he is aiming at when all these methods of Professor Webster are applied. Is it one of the aims of the policy to choose one of the systems?

R. Moss *(UK)* The aim of the co-operation between Professor Webster and the Department of Agriculture in Great Britain, and the Veterinary Service in particular, was simply to gain information. As far as we were concerned there was the possibility of using our own field staff in an observational study to obtain information on the behaviour of calves in the various systems. I discovered that Professor Webster had, in a very small way, a post-graduate student looking at this so we decided to co-operate together. We hope that we will be able to take certain decisions when we have all the information available, but that information is not available yet. I would not like to prejudge a particular system and say it is not going to be acceptable. I think Professor Webster has indicated the sort of thinking that may come out of the total study and the sort of decisions that may be recommended.

W. Sybesma I do think that it would be useful to think of alternative approaches of evaluating production systems. There might exist in other countries different approaches about which it is important to learn. Perhaps Dr. Andreae can tell us how the problem of evaluating different production systems is approached in Germany?

U. Andreae *(FRG)* That is too difficult to do at this particular moment.

G. van Putten *(Netherlands)* I think that this question has hung above us throughout this meeting. It concerns the conclusion that Professor Webster made at the end of his paper when he suggested that the individual rearing systems should be abandoned in favour of straw yards.

A.J. Webster *(UK)* I was very careful not to say that.

G. van Putten Yes, I noticed that you were very careful. Are you sure that these straw yards would solve most of the problems?

A.J. Webster Dr. Steenkamer is giving a paper later on the performance of veal calves in different systems and it might be appropriate for me to wait until then to discuss performance in different systems.

Specifically, I said that in our opinion there is no justification in terms of the health of the calf over 4 weeks of age for keeping it permanently isolated. I can expand upon this if necessary. In terms of the performance of the healthy calf, I have indicated already that the *ad libitum* teat feeding system is not as satisfactory as the bucket rearing system, commercially that is. Firstly, the food conversion ratio is slightly worse, possibly due in part to increased activity, but probably due more to the natural pattern of intake in the calf. This is too high at 8 weeks of age and too low at 16 weeks of age, whereas with buckets you can control this better. Therefore, I don't necessarily see the straw yard system, as we are using it, as the only answer. There may be an individual bucket feeding system in cubicles or individual feeding stalls which allows calves the ability to move around if they so wish. All I did say was that the housing of calves in individual crates, after the age of about 4 weeks, is done purely for the convenience of the producer. Before 4 weeks it can be justified to some degree on health grounds. This means that welfare is not taken into consideration at all and I think that it is important in discussing veal calves, pigs or battery hens, to remember that all the existing systems have been designed purely for convenience and economics. Although we should not try to abolish all the systems and substitute them with systems that we do not fully understand, we must evaluate, as a matter of philosophy, all these systems taking into account welfare (which, as I defined and according to the British Welfare Codes, is a system that meets both the health and the behavioural needs of the animals).

So, I do believe in the context of veal that we could evolve out of permanent crating but I also believe that it is too soon to do so. However, I certainly do believe, as you do, that a 60 cm wide crate is much too narrow.

G. van Putten I was not asking about performance figures, I was trying to talk about welfare.

A.J. Webster But you cannot neglect performance.

G. van Putten Of course, but we are here at the moment to try to find systems which are acceptable from both the performance and the welfare viewpoints, so we must compromise. I wanted to ask you whether in your view the straw yard would be the right compromise. I still have doubts about this just as you apparently have.

A.J. Webster One final point, the straw yard system is in its infancy in a scientific sense. There is a lot that we need to know about the performance, health and behaviour of calves in a straw yard before I would go as far as recommending this system. I do not necessarily suggest that it is the perfect solution, but we must continue to work towards finding this solution.

P.J. O'Connor *(Ireland)* Looking at the figure on the time spent on self-grooming, it is obvious that healthy calves will self-groom but this behaviour trait will disappear if the calves are about to become sick. We must provide sufficient space within crates for the calves so that they can efficiently groom themselves. The figure for the calves suckling their dams was lowest at 1.4 and I wanted to ask whether the dams continue to groom the calves.

A.J. Webster That is a very good point. There are two reasons why the calves with their dams do not groom themselves very much, relative to the calves in straw yards. The main reason is that most of the grooming is done by the dam herself. We do have figures for this but I did not include them in this paper. The second reason is that calves with their dams at pasture during the British winter retain their winter coats. A veal calf sheds its first coat which is something to do with its metabolic rate and so the veal calves at 10 weeks of age are shedding their birth coat and that is why they groom themselves a lot.

A.A. Jongebreur *(Netherlands)* Professor Webster, in the paper you gave the size of individual crates - the breadth was at least 0.8 m for veal calves. Previously, we have heard from Dr. Bogner and Dr. van Putten who gave different sizes. Perhaps it is worthwhile to begin with starting points from the behavioural side. Dr. Bogner said that the calf should at least have room to stretch its limbs. Dr. van Putten stated that in our conditions in the Netherlands it is not necessary to begin from that starting point. Dr.

Webster mentioned the starting point of 0.8 m, and the calves must have enough room to lie with their head on their ribs to fulfil their sleeping behaviour. I think it would be worthwhile to discuss these different starting points.

W. Sybesma But who will give the final answer - that's another question. Perhaps Dr. van Putten and Dr. Webster who have differing views can discuss some ideas about a possible compromise. Dr. van Putten, you have made clear why you have come to the figure of 70 cm, but what do you think of Dr. Webster's proposed 80 cm?

G. van Putten Mr. Chairman, I do not think 80 cm will allow the calves to lie on their sides. A young calf would need more than 80 cm and a calf ready for slaughter would need at least 105 cm. So 80 cm does not provide a solution to this problem of enabling the calf to lie on its side. I am convinced that, in the conditions found in northern Germany and Holland, if we tried to enforce a box width of more than 70 cm, this would not be accepted for economic reasons. We would then have to change to quite another system, maybe the straw yard although I am not happy with this system because it involves much labour and heat dissipation for the calves is not as easy in this system as in the slatted floor system. If the 70 cm crate is not acceptable from the point of view of welfare, then perhaps we should go as far as to begin thinking of abandoning the type of veal calf that we have at the moment, to think of changing over gradually to producing baby beef or something similar. However, I really do not think that the 80 cm box is an acceptable compromise.

A.J. Webster May I make a biblical point which has to do with the casting out of devils. The man cast one devil out of his house only to allow six more evil spirits in. Any form of legislation designed to improve welfare in casting out one devil is just as likely to let in a whole lot of new ones.

Supposing one attempted to ban veal production on a national or international basis. This would instantly mean there would be a vast number of calves surplus to the dairy industry and to the likely demands for beef, or the likely ability to support beef from cereals and grassland. In these circumstances the value of the calf becomes nothing and you get the 'Bobby' calf market. Here calves for two or three weeks of their lives are put from market to market, having no value, are slaughtered having no value and the insults to welfare on such calves are far, far greater than I am prepared to accept exists in any form of veal rearing system. In these, at least the calf has a value and so the farmer has an interest in keeping that calf alive. Any system that destroys the value of the calf will not be in the interest of calf welfare.

G. van Putten I have no desire to destroy the value of the calves, I was thinking of using them as meat producers but in quite different systems. I would never throw them away, that would be a waste in every sense, but I was suggesting changing to another system of beef production with the same calves.

W. Sybesma Dr. Fiems compared the suckling system with the bucket system and said that in some ways it was a better system but had higher costs. Could he elaborate a little on some details of these costs? I think this is important because although the suckling system may not be an alternative, it may give us an alternative when 'Bobby' calves appear.

L.O. Fiems *(Belgium)* If we take the cost of the feed intake of the dam into account with the feed cost of the calf, then we got a total feed cost of about 43 BF/kg liveweight gain for suckling calves compared to about 32 BF for bucket fed calves. A good question is whether it is right to take the whole feed intake of the dam into account and calculate it in the feed cost of the calf?

W. Sybesma But the cost is still higher? And this is per animal?

L.O. Fiems Yes, it is higher. In this figure the whole feed intake of the cow is calculated and then divided by the number of calves she suckled, expressed per kg liveweight gain. This is multiple sucking with an average of more than three calves per cow.

P. Le Neindre *(France)* It is difficult to say that because it is a very complicated problem. Cows with multiple sucklings do not have the same purpose as others. You must integrate in this exactly what a mother is, what she does throughout the year, how she can be managed. The pasture must also be integrated. I think it is very dangerous to give the price per kg of weight. One must be very careful with this multiple suckling. In France, no farmer wants to do this.

L.O. Fiems I agree with you. Multiple suckling in Belgium is also rather uncommon. We have seen suckling with Charolais or with our Belgian Blue White breed.

I would like to comment on what Dr. Steenkamer said about the behaviour of the calves who had three milk meals omitted over the weekend. I think that there is a good alternative to this system. If behaviour is an interaction between health, behaviour and performance, I think we can feed them only one milk meal per day. With this system the calves do not cry at all, performance is as good as with twice daily feeding, and once daily feeding stimulated starter intake. This also means that weaning is easier and the calves had fewer problems consuming solid feed.

N. Steenkamer *(Netherlands)* This is a compromise that I would like to suggest. I am not an ethologist, but I think that from a welfare point of view we must exclude the first few weeks of life in this system. The very young calf - one, two and three weeks of age - is a mammal and really should be fed a milk ration every day of the week, three times a day if possible, or if this is not possible in practice then twice a day. After a number of weeks then once a day feeding has in fact a direct advantage in that the intake of starter feed is increased which is good for economic and other reasons. However, to starve a very young calf is something I do not believe in.

L.O. Fiems I agree, and I must say that the calves in our trials were purchased when one and two weeks old. So they were in the third week of life when the trial started.

N. Steenkamer This is of course all part of behavioural observations. This is the topic of the Session, we are not talking about genealogy, blood composition, or the absorption of the feed from the small intestine. When you do this and analyse the blood of calves fed once or twice a day, whatever you find you will find this effect. I do not know how to translate this into welfare or comfort, but physiologically more frequent feeding is better. This is true for men and animals and, indeed, we could make do on one feed a day too. As long as economics allow for more frequent feeding, or even if economics do not allow for it, then it should be kept as good practice.

J.P. Signoret *(France)* From the conditions of free-living cattle, or mammals related to cattle, it can be seen that the maternal behaviour in such species could be divided in two types. There is the hider type and the follower type. The sheep is of the follower type - the mother segregates from the flock and is followed by the young up to the time when both the dam and the young go back to the flock. Cattle and goats are of the hider type - here the mother segregates for calving, then leaves the calves to return to the herd and has to move for suckling. It has been observed with animals who behave in such a way, that suckling occurs only twice or even once a day. So such suckling is not abnormal although I think the calves will suckle more frequently if possible. It could be that the calves are able to cope with reduced feeding rates because it is a remnant of what could occur in the natural situation. This does not mean, however, that the wild situation is best for welfare.

HOUSING VEAL CALVES
SOME RESULTS ON HOUSING SYSTEMS

J.P. Quillet
ITEB, Monvoisin, Cedex 71 bis, 35650 Le Rheu, France.

INTRODUCTION

The experiments reported here were done in an experimental station. However, the conditions were similar to those in practice: dairy calves were bought at 1 - 2 weeks of age and fed with a milk replacer for 15 - 16 weeks to produce 'white' calf meat.

The experiments were not designed primarily to study calf welfare but were aimed at developing practical production techniques. However, environmental conditions deal not only with welfare but also influence productivity. Even from the economic point of view, management has to take account of the well-being of the animals.

In general, environmental conditions were not studied alone, but in relation to other factors such as genotype, external characteristics of the calves, etc. Five successive experiments were designed to analyse the effect of genotype, sex and phenotypic characteristics of the calves at the beginning of the experiment.

In each case, two types of building were compared:

1) a fan ventilated, insulated building ('controlled environment'),
2) an uninsulated, naturally ventilated building ('climatic housing').

The aim was not to obtain complete climatic control, but rather to improve the environmental conditions with relatively simple systems suitable for use in commercial practice.

HOUSING (Figure 1)

'Controlled environment'

This consisted of a closed and insulated building, with

Fig. 1a. Climatic housing building and controlled environment building (behind).

Fig. 1b. Group penned calves at the end of the fattening period in the climatic housing building.

artificial heating, controlled humidity and fan ventilated. The calves were kept tied up on a slatted floor with individual head stalls.

'Climatic housing'

This was an uninsulated building; the south east side was open or fitted with a perforated inlet to reduce the wind speed. The calves were kept in group pens for eight animals with straw bedding and individual feeding of milk replacer.

ANIMALS

The calves were of three genetic types:

- Pure bred French Friesian
- Crossbred Friesian x Holstein (either 50 or 75% Holstein)
- Normand.

The percentage of females was never more than 25% and the genotypes were distributed evenly among the experimental groups and housing conditions.

RESULTS

Environmental variables

Temperature

Average daily maximum and minimum temperatures in the 'controlled environment' house are presented in Figure 2a. The values are relatively stable. There was one case of low temperature which was caused by a failure of the heating system, and the maximum rose also in one case because no cooling was applied due to the moderate ambient temperatures.

In the 'climatic housing' building (Figure 2b), the temperature range was much wider when compared with the 'controlled environment' both in the same week (10 - 12° vs. 5 - 6°C) and during the whole course of the experiment (25° vs. 11°C).

The average temperature was slightly lower in the 'climatic housing' than in the 'controlled environment' which is a normal result (Table 1).

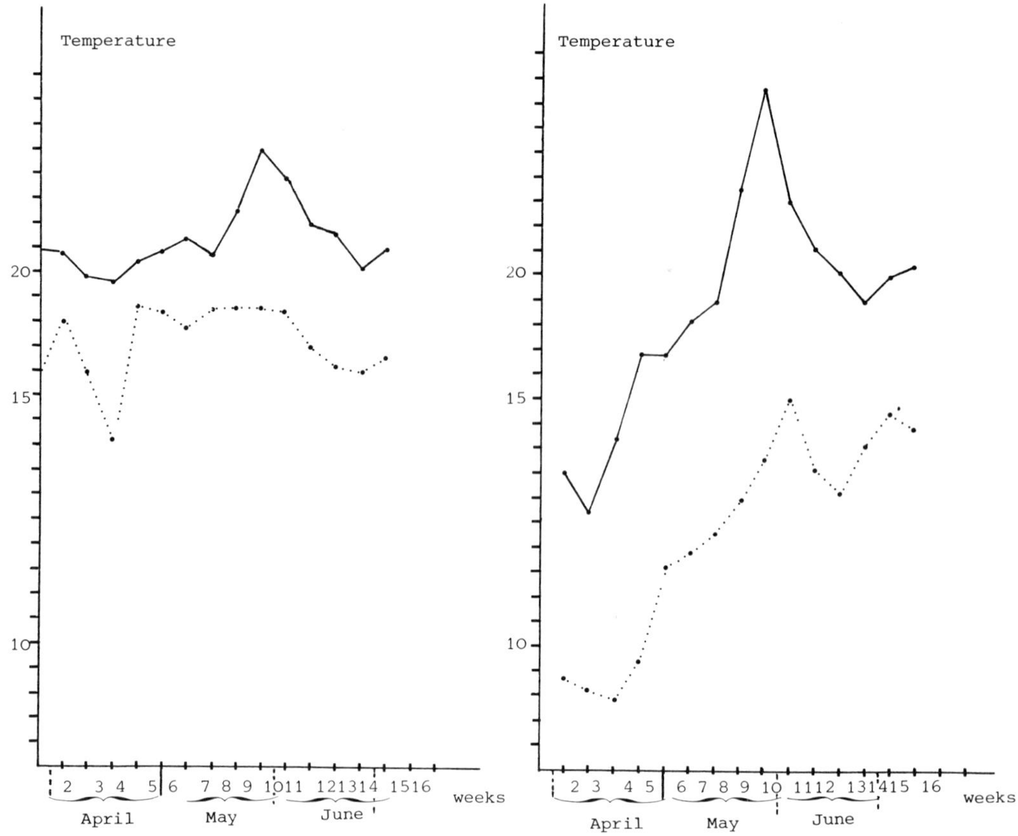

Fig. 2a. 'Controlled environment'. Fig. 2b. 'Climatic housing'.

Fig. 2. Mean maximum temperatures and mean minimum temperatures.

Relative humidity

In the 'controlled environment' building, the relative humidity was kept within a limited range by temperature control and artificial control of the humidity. In contrast, the range of variation was very wide in the 'climatic housing' (Table 1).

It can be concluded that the average temperature and relative humidity ranges are different between the two types of housing especially during the summer.

In the 'climatic housing' the animals are subjected to a wide range of temperature and relative humidity which raises the question of a possible adverse effect of such conditions.

TABLE 1

AVERAGE TEMPERATURES AND RELATIVE HUMIDITIES IN THE TWO HOUSING SYSTEMS

Housing	Temperature t (°C)	Δ t (°C)	RH (%)	Δ (RH) (%)
1) Trials No. 2 + 7: in winter				
Controlled environment	14	5 (M) 2 (m)	80	18 (M) 8 (m)
Climatic housing	9	12 6	85	55 25
2) Trials No. 6 + 8: in summer				
Controlled environment	20	12 3	74	43 16
Climatic housing	18	18 10	70	75 40

(M) = Maximum value of temperature (or RH) range/24 h
(m) = Minimum value of temperature (or RH) range/24 h

ANIMALS

General results

In the absence of reliable welfare indices that can be used under practical conditions, our observations are limited to the growth rate and the general health of the animals.

The average growth rate was not significantly influenced by the housing conditions (Table 2). Despite the differences observed in the climatic conditions, the growth rate and health of the animals in the two housing systems were, on average, equivalent. However a clear, but non significant, seasonal effect can be seen (Table 2) which is much greater than the difference between the types of building.

Initial weight

Table 3 shows that in the 'controlled housing' system, the initial weight has a marked influence on growth rate. When studied in the two types of building it was found that the building had no influence on growth but there was still a difference between animals of 'light' and 'heavy' initial weight (Table 4).

TABLE 2

AVERAGE GROWTH OVER 15 WEEKS (g/day)

Season	Winter	Summer	Overall
Number	75	135	210
'Controlled environment'	1 242	1 204	1 217
(Percentage)	(102.0)	(98.9)	(100)
Number	75	135	210
'Climatic housing'	1 238	1 202	1 215
(Percentage)	(101.7)	(98.8)	(99.8)
Difference (%)	0.3	0.1	0.2

TABLE 3

EFFECT OF INITIAL WEIGHT ON GROWTH (g/day)

Group			'Light'	'Heavy'
Initial body weight		(kg)	37.4	47.6
Weight at slaughter		(kg)	155.2	177.7
Daily weight gain (g/d)	0 - 56 d		934	1 036*
	56 - sl		1 339	1 470*
	0 - sl		1 122	1 238*
Respiratory disease up to 56 d.			11	2
Dead calves (no)			2	0

* Signif. ($P < 0.01$)

TABLE 4

INITIAL WEIGHT AND HOUSING

Group			'Light'	'Heavy'
Initial body weight		(kg)	41.1	48.2
Daily weight gain (g/d)	'Controlled environment'		934	961
	'Climatic housing'		944	964

Sex - genotype

A similar pattern emerges for the comparison between sexes and between genotypes (Tables 5 and 6).

TABLE 5

EFFECT OF SEX

Sex		Male	Female
Daily weight gain over 15 weeks (g/d)	'Controlled environment'	1 252	1 167
	'Climatic housing'	1 236	1 189

TABLE 6

EFFECT OF GENOTYPE

Genotype		French Friesian	Crossbred FF x Holstein		Pure Normand breed
			50% Holstein	75% Holstein	
Daily weight gain over 16 weeks (g/d)	(1) 'Controlled environment'	1 308	1 302	1 241	1 270
	(2) 'Climatic housing'	1 295	1 278	1 253	1 272
	$\frac{(1) - (2)}{(1)}$	0.99	1.84	0.97	0.16

Interactions between external temperature and housing

In winter, the comparison between the 'climatic housing' and the 'controlled environment' with or without artificial heating, showed no difference in any of the criteria measured (Table 7). Despite the fact that both types of housing were equivalent overall, the observations made during the summer of 1979 suggest that there was an adverse effect of high temperatures. A first period of relatively high temperature when the calves were 6 - 7 weeks old had no effect, but during a second hot phase, from 10 - 12 weeks, when feeding was very intensive, the higher temperatures in the 'climatic housing' system had a detrimental effect on the growth rate of the heaviest calves

TABLE 7

COMPARISON BETWEEN 'CLIMATIC HOUSING' AND 'CONTROLLED ENVIRONMENT' WITH OR WITHOUT ARTIFICIAL HEATING

Housing		'Climatic'	'Controlled environment'	
			Without artificial heating	With artificial heating
Initial weight	(kg)	43.8	43.3	43.3
Weight at slaughter	(kg)	164.9	163.0	161.7
Daily weight gain (g)	0-56 d	907	871	865
	56-sl	1 464	1 446	1 426
	0-sl	1 164	1 139	1 127
Feed conversion (kg/kg w. gain)		1.41	1.41	1.45
Diarrhoeas	(%)	77	50	77
Respiratory diseases up to 56 d	(%)	62	78	54

TABLE 8

GROWTH RATE AND DISEASES PERCENTAGES DURING THE WARM PERIOD

Sex and group		Female		Male 'light'		Male 'heavy'	
Building		'Controlled environment'	'Climatic housing'	'Controlled environment'	'Climatic housing'	'Controlled environment'	'Climatic housing'
Number of calves		16	16	14	13	31	30
Initial weight (kg)		41.8	42.0	40.9	41.3	48.4	48.0
Weight at slaughter (kg)		170.3	172.0	171.4	172.0	184.4	187.7
Daily weight gain (g)	56-sl	1 405[a]	1 433[a]	1 448[a]	1 441[a]	1 522[b]	1 420[a]
	0-sl	1 168[a]	1 181[a]	1 186[a]	1 188[a]	1 237[b]	1 188[a]
Digestive diseases f-56 d to sl (%)		31	19	19	44	34	53
Respiratory diseases f-56 d to sl (%)		12	37	19	12	6	3

(a) and (b) = (b) differs significantly from (a) (same line $P < 0.05$)

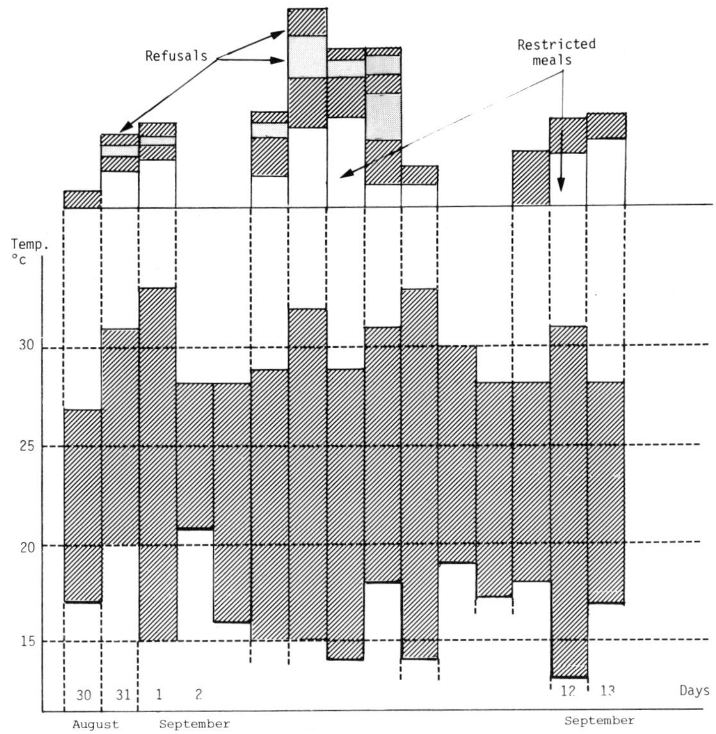

Maximum and minimum temperature and temperature range, every day, during the warm period

Fig. 3. Refusals during the warm weather - 'climatic housing'.

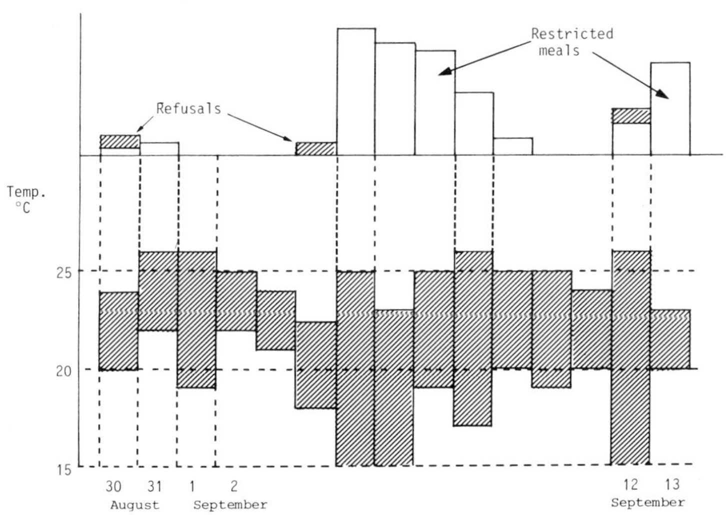

Maximum and minimum temperature and temperature range, every day, during the warm period

Fig. 4. Refusals during the warm weather - 'controlled environment'.

(Table 8). In this case, a high frequency of food refusal was associated with the high temperature and wide temperature range recorded in the 'climatic housing' system, whereas this was not observed during the same period in the 'controlled environment' system (Figures 3 and 4).

The rate of heat production per body surface unit, as calculated from metabolic data, results in an earlier handicap for heavier calves when they are subjected to higher temperatures.

CONCLUSIONS

Veal calf production from dairy cattle can take place in relatively simple buildings.

Contrary to the general opinion of the producers, low temperatures in winter generally have no major adverse effect, whereas even in the moderate climate of north western France, periods of heat can have a marked detrimental effect at least on growth rate.

A closed type of building is less sensitive to high temperatures when compared with an open one. Furthermore, the maximum temperature can only be controlled reliably in a closed building, with or without temperature control and air cooling.

Such results have to be considered in the design of calf housing, especially for intensive production.

APPENDIX

Relative Humidity (RH), Relative Humidity range $\Delta(RH)$, $\frac{\Delta(RH)}{\Delta t}$ ...., interaction between temperature and RH,.... are important parameters (possible explanation of respiratory diseases, for instance). They are not developed here in this short presentation.

DISCUSSION

N. Steenkamer *(Netherlands)* I was rather surprised at the temperature levels in the controlled environment versus the climatic house. If I remember correctly, in the house where no climatisation was used, the temperatures during the hot season were on average not higher but even lower than in the house where ventilators were installed. If one looks at the temperatures between 20 - 25°C, there are variations but there are many below the 20°C level. I would have thought this would be beneficial rather than harmful for the calves during this hot period. So why was there this negative effect for this group?

J.F. Grongnet *(France)* I am afraid that I do not know why this is so.

A.J. Webster *(UK)* What material was the roof of the climatic housing made of? Was it insulated?

J.F. Grongnet No, not in the climatic housing.

A.J. Webster Then in these circumstances the air temperature in the climatic house does not accurately describe the heat load on the calves. When the sun is shining on the roof of the climatic house, the radiant heat load on the calves will make them hotter than will be estimated from measuring air temperature.

EFFECT OF VARIOUS ENVIRONMENTS ON CALF PERFORMANCE

F.J. Harte and R.J. Fallon
Agricultural Institute,
Grange, Dunsany, Co. Meath, Ireland.

ABSTRACT

In Ireland the vast majority of calves are born in the spring resulting in a big demand for calf housing in that period of the year.

Climatic conditions in Ireland are not severe. There is only approximately 11°C difference between the mean mid-winter temperature and that of the mid-summer temperature. Wind speeds are relatively high.

Calves reared in different environments did not perform significantly differently from each other though there was a tendency towards increased calf performance in the solid sided pens compared to the open sided pens.

Our recommendations regarding calf housing in Ireland are:
1. *Calves should have dry lying.*
2. *They should have plenty of space - 2.5 m^2 and up to 7.5 m^3 in the house.*
3. *They should be free of draughts - air movement should not be greater than 0.3 m/sec, in the immediate vicinity of the calf.*

INTRODUCTION

In Ireland approximately 80% of all calves are born in the first half of the year and, indeed 50% of all these calves are born in the first three months of the year. All calves bred from dairy cows are removed from their dams one to three days after birth and are reared artificially. This means that a large amount of calf housing has to be provided in the first half of the year and is little used thereafter, resulting in a reluctance by farmers in many cases to invest a lot of money in calf housing. Many of the calves born to the beef herd (i.e., approximately 20% of the total calf crop) are born outdoors and suckle their dams for about 8 months (Drennan, 1976).

In any discussion on the effects of environment on calf performance the prevailing climate is important. It can be seen that the climatic conditions in Ireland are not severe (Table 1) according to figures documented by Rohen (1975). There is only 11°C difference between mean temperature in January and July. The difference between inland (Birr) and coastal regions (Valentia) is less than 3°C within any one

TABLE 1

MONTHLY VALUES OF MEAN DAILY MAXIMUM, MEAN DAILY MINIMUM AND MEAN DAILY AIR TEMPERATURE (°C) OVER THE 30 YEAR PERIOD 1931-1960.

Station	Height above *MSL	Mean daily values	Jan	Feb	Mar	Apr	May	June	July	Aug	Sept	Oct	Nov	Dec
Birr	70	Maximum	7.3	8.0	10.4	12.7	15.8	18.4	19.2	19.3	17.0	13.5	10.0	8.1
		Minimum	1.4	1.7	2.9	4.0	6.2	9.2	10.9	10.8	9.1	6.3	3.6	2.5
		Mean	4.4	4.9	6.7	8.4	11.0	13.8	15.1	15.1	13.1	9.9	6.8	5.3
Dublin	47	Maximum	7.6	8.2	10.4	12.7	15.4	18.4	19.6	19.4	17.3	13.9	10.3	8.4
		Minimum	1.3	1.6	2.6	3.7	5.9	9.1	11.0	10.6	8.9	6.3	3.8	2.6
		Mean	4.5	4.9	6.5	8.2	10.7	13.8	15.3	15.0	13.1	10.1	7.1	5.5
Valentia	9	Maximum	9.4	9.3	11.1	12.5	14.8	16.8	17.7	18.2	16.8	14.2	11.5	10.1
		Minimum	4.5	4.3	5.4	6.3	8.1	10.8	12.3	12.6	11.2	8.9	6.6	5.5
		Mean	7.0	6.8	8.3	9.4	11.5	13.8	15.0	15.4	14.0	11.6	9.1	7.8

* Mean sea level.

month (Table 1). The positions of the recording stations are shown in Figure 1. Rohen (1975) has also listed the monthly rainfalls (Table 2) and relative humidities (Table 3). The latter figures are high by European standards, however, and wind speeds in Ireland are also relatively high (Figure 1). The mean annual wind speed for inland regions is 4 m/sec and for coastal regions it is at least 6 m/sec. This helps to minimise the effects of high relative humidities in calf houses. Dodd (1978) suggests that the wind speeds and directions (Figures 1 and 2) are the predominant factors to be considered in ventilating climatic calf houses. In Ireland the predominant wind direction is from the south west (Figure 2) and Dodd (1978) suggests that this should be taken into account in the siting of a calf house.

Various experiments were carried out at Grange to determine the effect of house type and design on calf performance.

THE EFFECT OF DEGREE OF EXPOSURE ON CALF PERFORMANCE

Over two winter periods a total of 90 calves was used in three experiments. There were 30 calves in each experiment (10 calves per treatment). The calves were allotted at random (at approximately 7 days of age) to the following three treatments:

A. Covered area	maximum roof height of 3 m (Figure 3)
B. Part covered area	solid walls and three sides to a maximum roof height of 1.25 m (Figure 4)
C. Uncovered area	solid wall and three sides to a height of 1.25 m (Figure 5)

The bedded area in each treatment was similar at 1.6 m^2/calf. In addition, the calves on treatments B and C had a further uncovered yard area of 1.6 m^2/calf. The bedding used in all treatments was barley straw. An effort was made to keep the bedded areas dry in all three treatments by adding straw when necessary.

All calves were penned in groups, had water *ad libitum* and were fed similar amounts of milk replacer daily through individual teats (Harte and Fallon, 1981). In experiments 1 and 2 the calves on all treatments were offered concentrates *ad libitum*

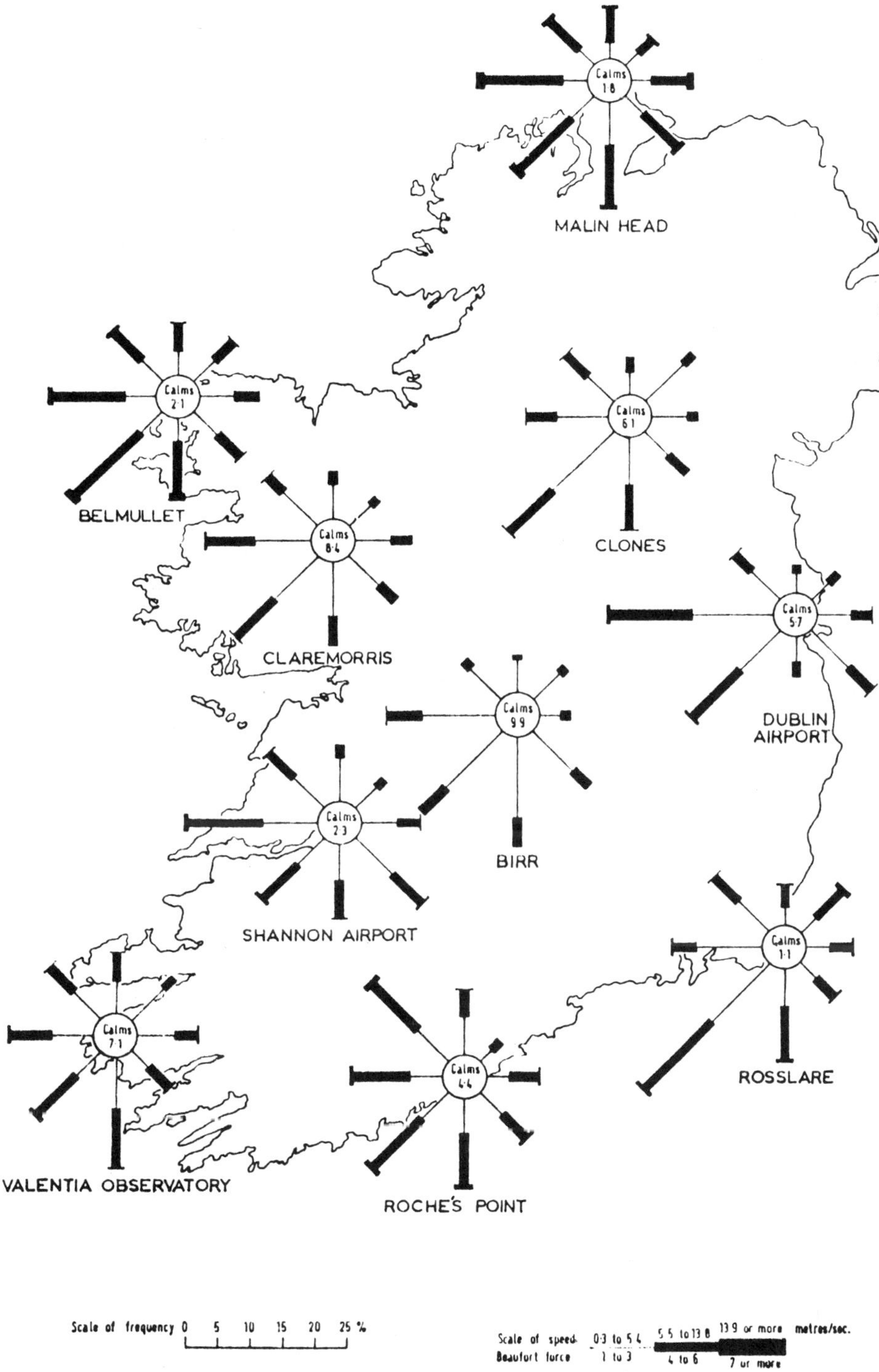

Fig. 1. Frequency of wind directions for groups of wind speeds at selected stations, 1961-1970.

TABLE 2

MONTHLY AND ANNUAL AVERAGES OF RAINFALL (IN mm) OVER THE 30 YEAR PERIOD 1931-1960

Station	Jan	Feb	Mar	Apr	May	June	July	Aug	Sept	Oct	Nov	Dec	Annual
Birr	85	59	52	49	62	60	86	82	87	84	78	89	873
Dublin	67	55	51	45	60	58	70	74	72	70	67	75	764
Valentia	167	107	103	74	86	81	107	95	122	140	151	168	1 398

TABLE 3

MONTHLY MEAN VALUES OF RELATIVE HUMIDITY (AS A PERCENTAGE) OF EACH OF THE HOURS 0h, 6 h, 12 h AND 18 h GMT* OVER THE 10 YEAR PERIOD 1961-1970

Station	Hours	Jan	Feb	Mar	Apr	May	June	July	Aug	Sept	Oct	Nov	Dec
Birr	0	90	88	88	89	87	89	89	91	91	90	90	90
	6	89	88	89	91	90	91	91	92	92	91	91	90
	12	86	80	74	70	68	68	71	71	74	79	83	86
	18	88	81	74	71	68	70	71	74	79	84	87	88
Dublin	0	85	84	83	85	85	85	87	89	88	88	86	86
	6	86	84	86	88	88	88	90	91	91	89	88	86
	12	82	77	72	71	69	70	70	72	74	77	80	81
	18	84	80	75	74	71	70	72	75	79	84	84	84
Valentia	0	83	81	83	84	85	88	89	89	88	86	83	83
	6	83	81	84	86	86	89	90	90	89	87	83	83
	12	80	76	74	75	73	76	78	79	79	79	79	80
	18	82	79	77	78	76	79	81	82	83	84	82	82

*Greenwich Mean Time

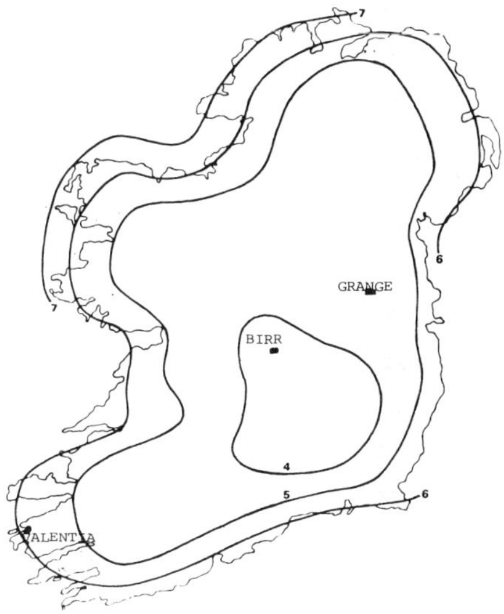

Fig. 2. Mean annual wind speed (m/sec), 1951-1970.

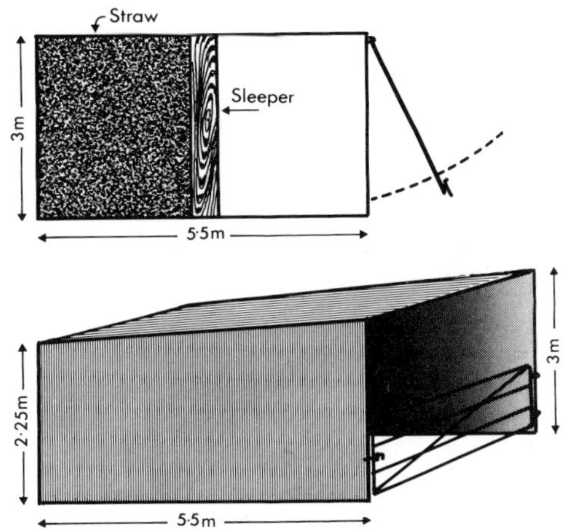

Fig. 3. Covered area.

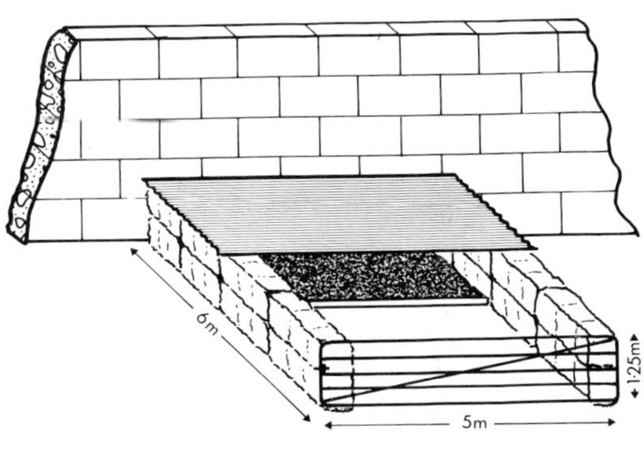

Fig. 4. Partly covered area.

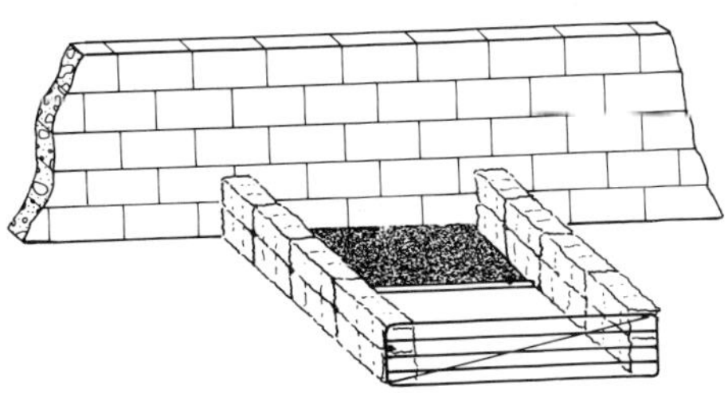

Fig. 5. Uncovered area.

while in experiment 3 the concentrates were restricted, such that each treatment group had the same total intake for the experimental period.

The experiments were carried out in December and January (Table 4) and the duration of each experiment was 42 days.

TABLE 4

CALF LIVEWEIGHT GAINS, FEED INPUTS (kg) AND FEED EFFICIENCIES (MJ/kg LIVEWEIGHT GAIN) IN EXPERIMENTS 1, 2 AND 3.

Experiment	Treatment	Gain 1-42 days (g/day)	Conc. (kg)	MJ/kg* gain
1	Covered area	470	16.7	29.9
	Part covered area	450	15.9	30.7
	Uncovered area	360	18.1	40.0
2	Covered area	480	20.5	31.4
	Part covered area	400	17.3	35.5
	Uncovered area	420	20.2	35.7
3	Covered area	430	12.5	30.0
	Part covered area	360	12.5	35.7
	Uncovered area	400	12.5	32.2

*1 kg milk replacer = 17.5 MJ   1 kg concentrates = 11.5 MJ

Mean initial liveweights of calves in Experiments 1, 2 and 3 were 42.4, 41.7 and 41.5 kg respectively.

Mean milk replacer intake per calf in all 3 experiments was 22.7 kg.

The starting dates for Experiments 1, 2 and 3 were January 13, December 7 and January 31 respectively.

RESULTS AND DISCUSSION

The calf performances for the three experiments are given in Table 4 and the climatic conditions prevailing during the experimental periods are summarised in Table 5. While liveweight differences were not significant between treatments in any of the three experiments, there was a tendency in all cases for those calves in the fully covered area to perform better than either of the other two treatments. However, performance by the calves on all treatments in all three experiments was satisfactory taking into account the levels of feed inputs used,

TABLE 5

SUMMARY OF DAILY CLIMATIC CONDITIONS DURING EXPERIMENTS 1 TO 3 (% OF DAYS)

Experiment	Minimum temperature °C	Mean relative humidity %	Mean rainfall (mm)	Wind speed (m/s)
	Below 5.0	Above 85	Above 0.1	Above 2.5
1	88	45	62	86
2	36	60	60	69
3	74	52	57	79

particularly milk replacer. The results suggest that the young calf is capable of withstanding average Irish winter weather conditions.

These findings support the work of Webster et al. (1976) who concluded that the prevailing environment within a climatic chamber had no effect on growth or feed conversion efficiency of young calves until air temperatures fell below 5°C. They emphasised, however, that it was important that the calves were dry, out of draughts and were eating normally. Appleman and Owen (1975) concluded that air temperatures between 0°C and 21°C did not interfere with calf performance provided that there were not excessive wind currents when cold conditions prevailed. Mitchell (1975) and Williams (1975) both working in Scotland, where mean winter temperatures are lower than in Ireland, also reported satisfactory performances where calves were reared in different types of climatic houses.

THE EFFECT OF PEN DESIGN ON CALF PERFORMANCE

While the experiments reported above suggest that environment had little effect on calf performance, it is conceivable that confining calves to pens in an open type shed could affect their performance. Two experiments involving a total of 40 calves were carried out to determine the effect of solid or open-sided individual pens on calf performance. There were 20 calves in each experiment (10 calves per treatment). The calves were allotted at random (at approximately 7 days of age) to the following two treatments:

D. solid sided individual pens (0.9 x 1.8 m)

E. open-sided individual pens (0.9 x 1.8 m)

The pens were erected in a Dutch type barn, roof height 6.5 m. The total area enclosing the 20 pens was surrounded by a perimeter wall (3 m distance from the pens) and consisting of bales to a height of 1 m (Figure 6). The bedding used was barley straw. All calves had water *ad libitum*, were fed similar quantities of milk replacer and had concentrates *ad libitum* throughout the 42 day experimental period.

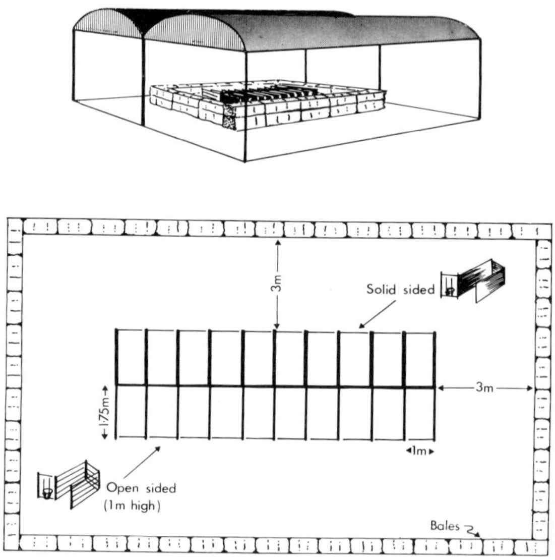

Fig. 6. Open and solid-sided pens in Dutch barn.

RESULTS AND DISCUSSION

Calf performance and feed intakes are given in Table 6. Table 7 shows a summary of the climatic conditions during the two experiments. Considerable variation occurred between calves within treatments but there was a tendency for the calves in the solid pens to have somewhat better liveweight gain and feed conversion efficiency. The level of milk fed and the concentrate intakes were low and this could affect performance in the relatively more exposed open-sided individual pens. In group pens the calves can select the optimal area to lie in as regards draughts and, of course, can shelter each other.

TABLE 6

THE EFFECT OF PEN TYPE WITHIN AN OPEN DUTCH BARN ON CALF PERFORMANCE

Pen type	Gain 1-42 days (g/day)	Milk rep. (kg)	Conc. (kg)	MJ/kg gain
Experiment 4				
Solid sides	420	20.0	15.2	29.5
Open sides	360	20.0	13.5	33.2
Experiment 5				
Solid sides	410	19.0	12.4	27.5
Open sides	300	19.0	6.8	32.9

The starting dates for Experiment 4 and 5 were November 14th and February 3rd respectively and the mean initial liveweight (kg) of calves was 41.6 kg.

TABLE 7

SUMMARY OF DAILY CLIMATIC CONDITIONS DURING EXPERIMENTS 4 AND 5 (% OF DAYS)

Experiment	Minimum temperature °C	Mean relative humidity (%)	Mean rainfall (mm)	Wind speed (m/s)
	Below 5.0	Above 85	Above 0.1	Above 2.5
4	65	50	67	76
5	74	26	48	76

EXPERIMENTAL WORK IN PROGRESS

Recently our colleague, Kelly (1980), carried out a number of experiments at Grange on different types of climatic calf houses. Four different types of houses have been compared and approximate details of these are tabulated in Table 8 where the number of air changes/house/day are also listed. Preliminary results are shown in Table 9 and there appears to be no significant difference between the various house types as regards liveweight performance. Results obtained as regards level of lung damage in calves and bacterial levels in the atmosphere are at present being analysed.

TABLE 8

SPECIFICATION OF THE FOUR HOUSE TYPES

House type	Number of calves	Number of compartments	Area/calf Floor m^2	Pen m^2	Cubic m^3	Inlet m^2	Outlet m^2	Relative number of air changes/ house/ day*
F. Calf kennel	32	1	2.89	1.25	7.17	0.080	0.093	100
G. Climatic	40	2	2.16	1.55	7.38	0.039	0.048	46
H. Fan-insulated	40	2	2.16	1.55	7.38	adjust-able	adjust-able	36
I. Converted cow byre	32	1	3.66	2.47	9.54	0.112		31

* calculated from the time taken for 1 smoke pellet to diffuse out of the house.

TABLE 9

THE EFFECT OF HOUSE TYPE ON CALF LIVEWEIGHT GAIN AND FEED EFFICIENCY

House type	Number of calves	Gain 1-42 days (g/day)	Milk rep. (kg)	MJ/kg gain
Experiment 6				
F. Calf kennel	16	440	33.0	31.2
G. Climatic	16	450	32.9	30.5
H. Fan insulated	16	420	32.6	31.3
I. Converted cow byre	16	430	33.0	31.6
Experiment 7				
F. Calf kennel	24	430	31.8	30.5
G. Climatic	24	380	31.7	34.8
H. Fan insulated	24	400	31.4	32.5
I. Converted cow byre	24	410	31.6	31.8

Mean initial liveweights of calves in Experiments 6 and 7 were 42.9 and 45.6 kg respectively.

The starting dates for Experiments 6 and 7 were January 18 and February 25 respectively.

CONCLUSIONS

As a result of the various experiments reported here, it is concluded that in the Irish climate, temperature is not likely to inhibit calf performance. However, if calves are individually penned, then it would appear that draughts, and perhaps temperatures, can be a problem. Our recommendations therefore as regards calf housing in Ireland are as follows:

1. Calves should have dry lying - the young calf lies down up to 80% of its early life and the surface with which he is in contact should be dry.
2. Calves should have plenty of air space to allow for air movement - each calf should have at least 2.5 m^2 and up to 7.5 m^3 in the house.
3. Calves should be free of draughts - air movement in the immediate vicinity of the calf should not be greater than 0.3 m/sec.

Finally we would like to say that problems such as the high relative humidities may not be a problem in many other European countries. We must emphasise that mortality in any single-birth species of animals is a major problem and so the general health is very important. Factors such as the optimal number of calves per air space, frequency of occupation and the actual siting of buildings relative to wind direction need further research.

REFERENCES

Appleman, R.D. and Owen, G.F., 1975. Breeding, housing and feeding management. J. Dairy Sci. 58, 447-464.

Dodd, V.A., 1978. The essentials of calf housing. In: Progressive Farmer. Vol. No. 2, No. 4 (July) p. 20-21, Dublin.

Drennan, M.J., 1976. Beef from the suckler herd. Hand Book, No. 9, An Foras Taluntais, Dublin.

Harte, F.J. and Fallon, R.J., 1981. Calf rearing. In: Calf to Beef Handbook, No. 17, An Foras Taluntais, Dublin.

Kelly, T.G., Tuite, P.J., Fallon, R.J., Dodd, V.A., Ruane, D. and Harte, F.J., 1980. Effect of house type on calf performance and health. Animal Production, Research Report, An Foras Taluntais, Dublin.

Michell, C.D., 1975. Climatic calf housing. A study of the feasibility of naturally ventilated calf housing. Ph.D. Thesis, University of Aberdeen.

Rohan, P.K., 1975. The climate of Ireland. The Stationary Office, Dublin.

Webster, A.J.F., Gordon, J.G. and Smith, J.S., 1976. Energy exchanges of veal calves in relation to body weight, food intake and air temperature. Anim. Prod. 23, 35-42.

Williams, P.E.V., 1975. An examination of climatic housing and nutrition on the performance of calves. Ph.D. Thesis, University of Glasgow.

DISCUSSION

N. Steenkamer *(Netherlands)* What is the availability of straw in Ireland and what is its price? We have seen slides of calves lying happily in straw, but how much would this cost in Ireland?

F.J. Harte *(Ireland)* The cost of straw *per se* out in the field is not very much, but the handling is another matter. We would certainly like to do away with straw and from a managerial point of view I would much prefer to put calves on slats. However, I still maintain that even that sort of calf kept on a completely slatted area will have a very wet floor under him in our climate, with quite a lot of humidity coming up.

H.C. Adler *(Denmark)* It looked from one of your slides as if many of your animals had bursitis which is a swelling of the fore knees. This is important since if all the animals have bursitis then they obviously have insufficient welfare.

F.J. Harte This is difficult for me to answer because I am not a vet. They are outdoor calves and are interesting because we had tried to look at the lactation of three cows and feed calves according to lactation. We fed four calves for a period of about eight weeks, then weaned them, then put on six more, and so on. These 64 calves were a group which were fed outdoors and so certainly faced no problems of standing on hard ground.

J.Y. Blom *(Denmark)* I must extend Dr. Adler's question and ask where were the feeding troughs placed? Are they placed on a concrete base at the corner of the field or just on grass?

F.J. Harte In fact, these calves were not fed concentrates, they were suckled once or twice a day depending on the number on per cow. The cows went into a 'shooter' race which might have been muddy but never on hard ground.

PRODUCTION RESPONSES OF YOUNG CALVES, FATTENING BULLS, AND REPLACEMENT STOCK TO DIFFERENT TYPES OF HOUSING

V. Jensen and S.P. Konggaard
National Institute of Animal Science,
Trollesminde, Roskildevej 48, DK-3500 Hillerød, Denmark.

INTRODUCTION

In Denmark milk production is becoming more and more concentrated on large scale specialised dairy farms. With an increased herd size and an increased number of cows kept in loose housing it has become more and more common to keep the calves in separate buildings. This has, however, revealed several problems, mainly because the climatic conditions in these houses are very poor because ventilation is insufficient, given the heat produced by young calves, to keep down the air humidity. It is possible to control this climatic factor by installing supplemental heating or dehydrating equipment, but with the exorbitant rise in oil prices the economy of such an installation is very doubtful.

An alternative is the traditional system where the entire herd is housed together. Between 1971 - 77 the Department conducted animal performance trials with calves, fattening bulls, and young stock in three different types of barns where this system was applied.

The trials were carried out at the National Experimental Farm Trollesminde, Hillerød, in three buildings identical on the exterior, but arranged inside as follows:

1. Insulated tie stall barn
2. Insulated free stall barn
3. Uninsulated bedded manure pack barn

Each building was an independent unit where replacement stock and fattening animals were kept under the same conditions as the milking cows. The number of animals kept in each unit was 25 dairy cows + followers. The parameters studied.were daily gain, feed intake, feed conversion rate, reproduction observation, breeding, health care and general management. For the fattening animals the carcase composition and meat quality were registered as well.

The animals in the different units received the same treatment and feed ration at all times, although there was a difference in that the young stock in the tie barn were fed individually and the animals in loose housing were fed on group basis. The calves in each unit were kept in individual straw bedded pens until they were about 3 months old. Then the animals in the tie barn unit were tied up and the animals in loose housings were put into groups of 4 - 8 animals on slatted floor compartments in the free stall unit and deep bedding compartments in the manure pack unit.

REPLACEMENT STOCK

At the beginning of the experiment the calvings were very concentrated as all the cows calved between 1st August and 1st January. Liveweight gain is therefore naturally divided into periods housed and periods on pasture until calving at about 25 months of age.

TABLE 1

GROWTH RATE OF YOUNG STOCK IN RELATION TO TYPE OF HOUSING

Daily gain, g	Tie stall (n = 60-33)	Slatted floor (n = 59-31)	Deep bedding (n = 73-37)
1st winter housed	698	693	691
1st summer on pasture	528	552	533
2nd winter housed	652	649	650
2nd summer on pasture	548	584	575
Birth to calving	669	653	661
Age at calving (months)	25.7	25.2	26.3
Weight at calving (kg)	550	533	566

It appears from the Table that the differences in growth rate were very small for the heifers from the three different types of housing. There seems to be a tendency towards a slightly better growth rate for the heifers in the loose housing systems during the summer period. The explanation may be that these animals were in better condition when they were put out to pasture and consequently were less affected by the change.

The heifers from the slatted floor unit calved at a slightly lower age and weight, but this was due to better pregnancy results (Table 3).

The conclusion must be that all three types of barn supplied conditions for a satisfactory growth rate during the rearing period.

FEED CONVERSION

Feed consumption per kg gain was calculated for the period during which the animals were kept in the respective barns on the basis of daily recordings of the allocated feed.

TABLE 2

FEED CONVERSION OF YOUNG STOCK IN RELATION TO TYPE OF HOUSING

FU per kg gain	Tie stall (n = 60-33)	Slatted floor (n = 59-31)	Deep bedding (n = 73-37)
1st winter housed	3.57	3.72	3.43
2nd winter housed	6.82	7.08	6.82

In the first winter period during which the animals were kept in individual pens for about half the period, the calves on slatted floor required most feed per kg gain and those on deep bedding least. In the second winter period the conversion rate was again better for the individually fed heifers in the tie barn and the deep bedding. However, the differences were not statistically significant ($P > 0.05$).

REPRODUCTIVE PERFORMANCE

The heifers were served at the first heat after 14 months of age, and artificial insemination was used for all the animals. The reproductive performance is shown in Table 3.

The Table indicates the same picture as that for the cows, i.e. that heifers kept in loose housing are easier to get in calf than animals kept in tie barns. On average the results were best for heifers kept on slatted floor, which seems surprising from a practical point of view. The explanation is likely to be found in the fact that it is easy to record animals

TABLE 3

REPRODUCTIVE PERFORMANCE IN RELATION TO TYPE OF HOUSING

	Tie stall	Slatted floor	Deep bedding
Number inseminated	38	35	38
Number pregnant	33	31	37
% pregnant	87	89	97
Number of inseminations per pregnancy	2.09	1.58	1.76
% pregnant after first insemination	36	65	51

in heat when they are housed in relatively small groups with ample space (2.5 m^2/animal).

FATTENING CALVES AND YOUNG BULLS

The experiments with fattening animals were started in a period when there was a market for the so-called skim milk calves. The animals were therefore slaughtered as fattening calves (liveweight 210 - 240 kg) and young bulls (liveweight 430 - 470 kg), respectively (Table 4).

TABLE 4

PRODUCTION RESULT FROM FATTENING MALE ANIMALS IN RELATION TO TYPE OF HOUSING

	Fattening calves			Young bulls		
	Tie stall	Slatted floor	Deep bedding	Tie stall	Slatted floor	Deep bedding
Number of animals	28	31	26	12	11	10
Liveweight at slaughter, kg	237	217	235	456	463	460
Daily gain, g	978	941	959	1 085	1 035	1 077
FU per kg gain	3.55	3.40	3.50	4.37	4.60	4.51
Dressing percentage	55.2	55.5	55.4	55.5	54.6	55.0
Classification (1-11)	6.0	6.3	5.8	7.4	8.5	7.8

There were only small and not significant differences in growth rate, feed conversion, dressing percentage, and meat quality for fattening calves. The slightly lower daily growth rate and the more moderate feed consumption for the calves on slatted floors can be related to lower slaughter weight. The growth rate for the young bulls is similar in the three types of barn. The feed conversion rate is slightly better for the animals in the tie barn, which is in accordance with results found in other experiments where tied bulls had approximately 5% better feed conversion rate. There was a tendency to better meat classification for the bulls on slatted floor. The differences, however, are not statistically significant ($P > 0.05$).

CALF MORTALITY

The rate of calves stillborn was rather high (average 8.9%), but the mortality after birth, which is directly related to the environment, was on the other hand very low (Table 5).

TABLE 5

CALVES BORN AND CALF MORTALITY 1971 - 75.

	Tie stall	Slatted floor	Deep bedding
Calves born	164	184	200
Stillborn	14	14	21
% stillborn	8.5	7.6	10.5
Dead within 3 days	2	4	0
% dead within 3 days	1.2	2.2	0
Dead after 3 days of age	3	1	5
% dead after 3 days of age	1.8	0.5	2.5

HEALTH

Apart from an outbreak of coli infection the health situation was satisfactory (Table 6).

The Table indicates that the health situation concerning respiratory diseases, leg and hoof problems, was better in the loose housing units. It is particularly interesting to notice that there was only half the number of respiratory diseases in the uninsulated deep bedding barn compared to the insulated tie stall and that no leg and hoof diseases occurred in this unit.

TABLE 6

DISEASE FREQUENCY (CASES PER 100 ANIMALS)

	Tie stall	Slatted floor	Deep bedding
Gastro-enteritis	50	55	46
Respiratory diseases	15	9	7
Leg and hoof diseases	7	3	0
Other diseases	7	7	8
Total	79	75	61

CONCLUSION

The overall results indicate that the calves and young stock perform satisfactorily and equally well in any of the three common types of housing, where the entire herd is kept together under one roof. Separate houses for the young calves does not seem to be necessary, particularly for small or medium size herds, and from an energy point of view they are also more expensive to operate. Building expenses for a highly insulated separate calf house with all the necessary climate control equipment are high. The daily cost for electricity and oil add considerably to the rearing cost, an expenditure which can be kept at a minimum if the calves are kept together with the cows. The young calves ought, however, to be placed in the warmest area of the house away from windows and doors which are frequently opened. For the first 2 - 3 months they should be kept in individual pens, which are easy to keep clean, and where the daily feeding and management will require a minimum of labour. If these requirements are met the production results can be expected to be as good for calves housed under the conditions mentioned as for calves kept in separate houses.

DISCUSSION

N. Steenkamer *(Netherlands)* Your statement was that in general the feed conversion ratio with the tethered beef animals was 5% more favourable in comparison with the other two treatments. You thought that energy or heat losses might account for this, but do you think this is really a satisfactory explanation for this phenomenon of 5%? This represents a great deal of money.

V. Jensen *(Denmark)* We do have figures for about 1 200 animals.

N. Steenkamer Therefore, with such a large number of animals, 5% really is a lot and it should be worth studying where it comes from.

V. Jensen We do believe that this comes from the calves running around. Physical activity could account for up to 5%, though no more.

L.O. Fiems *(Belgium)* At our Institute with half-year bulls of a liveweight interval of 250 - 550 kg we encountered the same phenomenon. Tied bulls had lower feed conversion, but I have no explanation for this.

V. Jensen You had the same 5%, or something similar?

L.O. Fiems Yes, about 4% better feed conversion.

H.C. Adler *(Denmark)* You mentioned that you had no lesions at all on the legs and hooves in the deep litter system. You had quite a few in the other two systems, the tied system and the slatted floor system. Were these severe lesions?

V. Jensen I cannot answer you because I was not there. I think a veterinarian was involved so it might have been rather severe.

W. Goldhorn *(CEC)* When talking about economics we usually have in mind the cost of manpower. However, oil prices will become more and more important in the economics. Perhaps there are possibilities for good insulation or even heat exchange by heating the incoming air by the outgoing air. There are some systems which could be used to do this. The higher the oil prices rise, the more important straw bedding will become. Almost all animals on straw bedding are better adapted to temperature changes and even humidity. The question to everyone here is what are the insurmountable disadvantages of straw bedding compared to alternative systems of veal production? You say that bucket feeding is good for feed conversion, well I know of producers who keep calves on straw but feed them from buckets. This combines the advantages of both systems and I really do not see exactly what is the great disadvantage of straw.

V. Jensen That is my personal opinion as well. If the calves are kept on straw then practically no housing is needed, only shelter. It is not temperature so much as the air humidity which is the problem in this type of closed in system. It is important to have enough space. One farmer has put newborn calves in the straw barn and never suffered from high mortality rates. Now that the calves are kept in central heating, temperatures controlled by thermostats etc., then mortality can be as high as 20%. I think we should go back to something more simple.

J.F. Harte *(Ireland)* I do agree that straw would be the simpler method and probably better for welfare too, but I think we are talking about something quite different. Take as an extreme case a farmer who has 200 cattle to feed and look after in the winter, he has to face not only the high cost of labour but also the unreliability of labour. I think this is the real problem and if you keep cattle on slats then they do not have to be bedded at the weekends, the farmer can do most of the work himself. This is why there are things like colostrum removers on milking machines, they are high cost operations (and indeed slats is a high energy wasting system due to the handling of slurry) but they remove the farmer's dependence on unreliable labour. The farmer wants to be independent of the worker who will not come in on Saturdays and Sundays.

J.M. Bienfait *(Belgium)* The major problem with straw bedding is straw availability. If you are going to pick straw you are going to need in Holland a surface exceeding that existing with cereals to provide the straw. Straw availability is a major problem.

W. Sybesma *(Netherlands)* I would like to sum up the last few papers. Dr. Fiems talked mainly about the economics and health aspects of different rearing systems. Dr. Webster drew our attention to the importance of methodology, especially in order to evaluate different systems. We heard from Dr. Quillet's paper that the importance of climate is not always as great as we might expect. Dr. Harte provided many details about the different systems used in Ireland. Finally, Dr. Konggaard told us about the new developments in housing systems found in Denmark.

I drew a few conclusions from these papers. In this wide range of systems presented to us, we need to make careful comparisons by multidisciplinary co-operation. This should involve behaviour, methodology, economics, aspects of veterinary science and, as we heard in the last paper, energy is becoming more and more important. So the conclusion must be that we should intensify the evaluation of the different production systems and then it might be revealed that the evolvement out of crates is not impossible. As to studies on crate systems, I think it is necessary to find an answer to what will be the minimum acceptable welfare condition. This generally refers to crate sizes - 70 cm, 80 cm or more. This makes clear that we must compromise.

GROUP HOUSING SYSTEMS FOR VEAL CALVES

A.A. Jongebreur[1] and A.C. Smits[2]
[1]Department of Agricultural Engineering, Agricultural University,
[2]Institute of Agricultural Engineering, Wageningen, Netherlands.

INTRODUCTION

Individual housing of veal calves is generally practised in the Netherlands. This system has advantages from the point of view of feeding routine, health inspection of the calves and prevention of possible mutual contamination of the calves. With regard to the welfare of the calves there are drawbacks, namely reduced freedom of movement, especially in getting up and lying down, the impossibility of turning round and the impossibility of certain lying positions (Jongebreur and Zwakenberg, 1976; Graf et al., 1976). These disadvantages are the more important as the slaughter weights of the veal calves have increased.

On the basis of research carried out by van Putten and Elshof (1974, 1980) and Jongebreur and Smits (1978), it can be stated that an improvement in the individual system can be made by adapting the dimensions of the boxes to a minimum size of 0.70 m x 1.70 m. We have however, to keep in mind that this increase in the size of the box does on a practical level result in higher building costs.

GROUP HOUSING

In the past decades experiments have been carried out with milk dispensing machines and group housing. This combined husbandry system resulted in a lot of problems, namely a relatively high feed conversion rate, special attention during the first weeks of production, urine drinking, navel sucking, more difficult health inspection and large variation in the daily gain (Krippl and Burgstaller, 1970). This unfavourable combination of housing system and feeding system removed group housing of veal calves from Dutch agricultural practice ten years ago. The increasing size of individual crates, together with the advantages of group housing in relation to animal welfare, were reasons for intensifying research in the field of group housing of veal calves. Attempts to form a group box out of three individual boxes, however logical and worthwhile the approach, were not successful due to the frequency of urine drinking and the problem of tying the calves.

RESEARCH

Research was set up in close co-operation with a milk replacer firm to compare a group housing system with a concrete slatted floor and individual boxes. Both penning systems were in one calf house and they are summarised in Table 1.

TABLE 1

CHARACTERISTICS OF THE GROUP HOUSING AND INDIVIDUAL PENS

Housing system	Group	Individual
Number of calves/box	5	1
Sizes of the box (m)	3.8 by 2.15	0.65 by 1.65
m^2/calf	1.63	1.07
Floor	reinforced concrete slats with straw; gap width 5-10 mm	wooden slats; gap width 15-25 mm
Number of calf places	20	19

Wheat straw was applied to the floor of the group boxes and barley straw was applied in baskets. The calves in the individual boxes were able to eat straw from the age of 6 weeks in the first trial and from the age of 2 weeks in the next trials. The results of four successive trials are known at present.

During three trials observations were carried out on the calves at 6, 12 and 18 weeks of age concerning several aspects of behaviour such as lying and lying positions, straw intake and ruminating, licking the stall equipment and self-licking. The different activities were recorded with the help of codes during periods of 24 h. The preliminary results of these observations are mentioned.

RESULTS

The technical results of the four trials concerning growth, feed conversion rate and killing-out percentage and straw intake are given in Table 2.

TABLE 2

AVERAGE PRODUCTION CHARACTERISTICS OF VEAL CALVES IN GROUP BOXES AND INDIVIDUAL BOXES IN THE PERIOD 0-21 WEEKS

Housing system	Individual		Group	
	Average	Variation	Average	Variation
Number of calves	76	76	80	80
growth: in kg	159.6	156.6-163.1	163.4	155.1-168.3
in percentages	100	-	102.4	95.1-107.0
feed conversion rate	1.71	1.59-1.85	1.68	1.59-1.82
in percentages	100	-	98.2	93.0-105.2
Mortality in number of calves	2	0-1	6	0-3
killing-out percentage	63.2	62.7-63.4	62.6	62.2-63.0
straw intake in kg per calf*	7.5	6.9-9.3	27	17.0-32.5
lying straw in kg per calf	-	-	115	81.0-149

* included the waste of straw during the intake.

The results in the third trial were quite different from the results in the other three experiments. These less good technical results for the calves kept in the group boxes were due to quite a number of animals drinking urine. In the other trials this urine drinking was observed in a restricted number of calves and could be controlled by the use of a muzzle for some weeks. Calves, which drank urine over a long period have been eliminated in reckoning the results.

Tables 3 and 4 give the preliminary results of the observations on the behaviour of the calves.

This experiment is still being conducted so that no final conclusions with regard to the production characteristics and the behavioural aspects of both systems are drawn. Health aspects are referred to in the paper of Dr. Steenkamer.

TABLE 3

LYING BEHAVIOUR IN PERCENTAGES OF A 24 h PERIOD OF VEAL CALVES KEPT IN GROUP BOXES AND INDIVIDUAL BOXES

Housing system	Individual	Group
Number of 24 h observations	10	10
Lying position:		
on the sternum	44.7	40.7
one or two hindlegs from the body	10.3	12.6
with stretched limbs	0.8	0.9

TABLE 4

AVERAGE STRAW INTAKE AND TIME (IN PERCENTAGES OF A 24 h PERIOD) USED FOR STRAW INTAKE AND RUMINATING

Housing system	Individual	Group
Average straw intake in g/calf/day	51	183
Average time used for straw intake in percentages	2.2	5.5
Ruminating (%)	11.3	16.5
Self licking (%)	2.5	2.6
Licking round the nose	1.3	0.7

These preliminary results show a strong similarity with the results of Graf et al. (1976).

DISCUSSION

In spite of the relatively satisfactory results of the group box with slatted floor and straw, there is still the question of the implications of this system for management practices of veal calf producers. It is quite clear that the use of straw has a high labour requirement because both bringing-in of straw and removal of manure containing straw are done by hand. Tying the calves at the beginning of the production period also requires extra labour. The relatively high labour requirement of the group box with straw means that it is not attractive at the moment in terms of practical application, although mechanisation of removal of manure containing straw may lead to a housing system for veal calves which offers a wider perspective for the future.

FURTHER RESEARCH

A more comprehensive study of several types of group housing, in close co-operation with the milk replacer firm 'Denkavit Netherlands Ltd.' and the Department of Animal Husbandry of the Agricultural University is being carried out.

As there is an overcapacity in the number of available veal calf facilities in the Netherlands, one has to take into account that the group pens have to fit the existing calf houses and therefore can measure a maximum of 2.4 - 2.7 m.

The various types of group boxes, most without straw on the floor, are mentioned in Table 5.

TABLE 5

TYPES OF GROUP PENS TAKEN UP IN A COMPARISON WITH THE COMMON INDIVIDUAL PENS

Type of group pen	Length of feeding fence in m/calf	m^2 per calf
Total slotted concrete floor with:		
1. separate lying area (Figure 1)	0.5	1.20
2. with straw during six weeks after the start of production	0.5	1.20
3. with straw during the whole production	0.5	1.20
Slatted floor of wood and concrete with:		
1. feeding partitions of 0.8 m length (Figure 2)	0.6	1.44
2. feeding partitions of 1.0 m length	0.65	1.56
3. individual boxes of 1.4 m length	0.65	1.56

The individual boxes have a size of 0.7 by 1.7 m and thus 1.2 m^2 per calf. In this research the group housing systems are evaluated both from the farm buildings' side as well as from the technical, ethological and economical point of view.

The first preliminary impressions give a slight preference to the slatted floor and a separate lying area. In this system

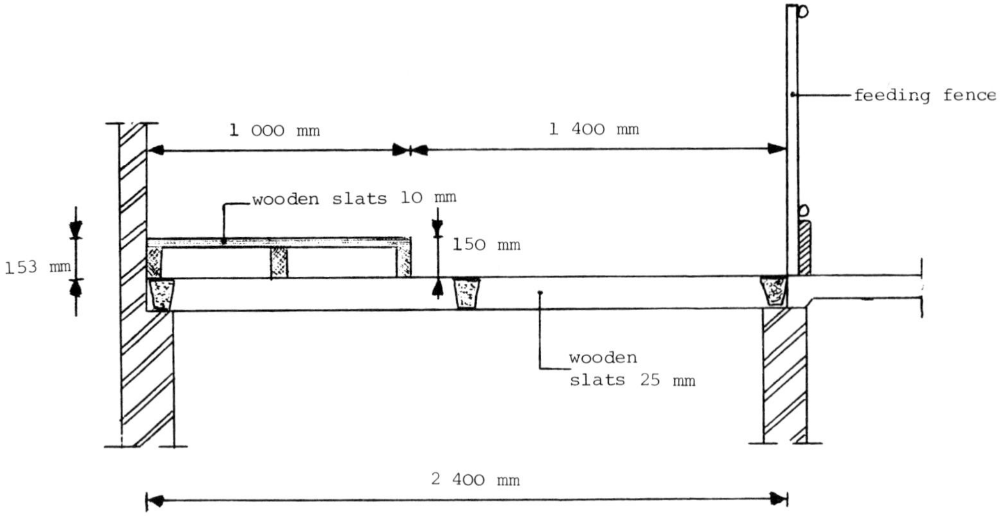

Fig. 1. Group pen with slatted floor and separate lying area.

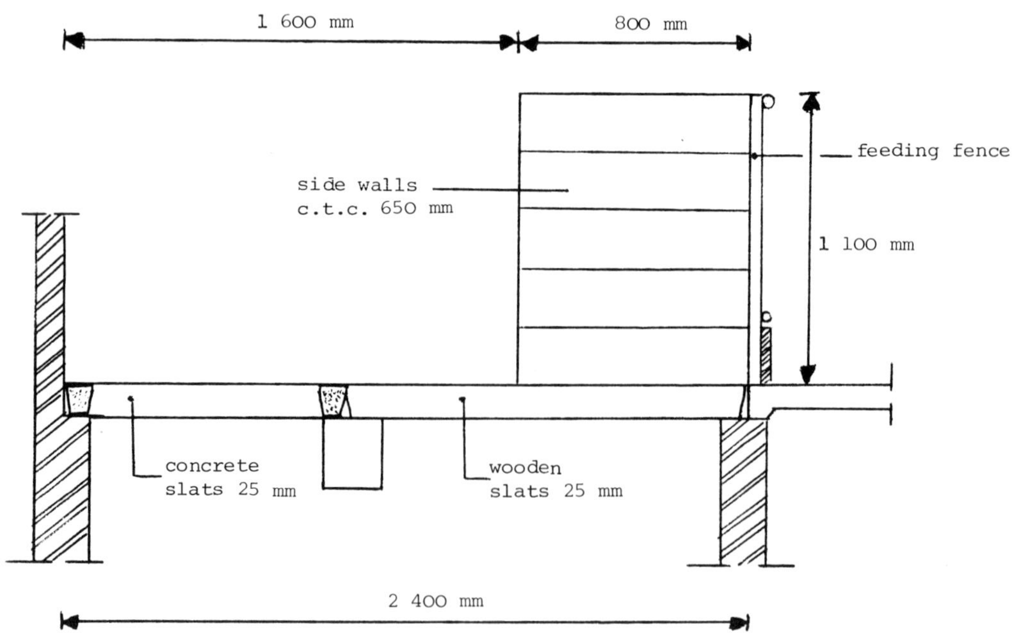

Fig. 2. Group pen with partly enclosed boxes.

no straw is used. Nevertheless, problems with standing up by the calves and a slippery floor did not occur (Graf, 1976). The improved consistency of the manure is apparently the reason for this phenomenon.

According to Bogner (1980) the space requirement of group housed veal calves with liveweights over 200 kg is 1.45 m^2. This is based upon the lying position with stretched limbs. One should raise the question if this is - in comparison with the minimum size of individual pens - a realistic criterion. If the starting point for group pens is a length of the feeding fence of 0.6 m in the same calf house - with individual pens of 0.7 m width - about 16% more veal calves can be kept as long as sufficient air space is available.

FINAL REMARKS

The preliminary outcome of the experiments with a group housing system with the use of straw, in comparison with the usual individual boxes, are promising with regard to production characteristics and animal welfare. The extra labour required, however, for this housing system is too great and therefore requires mechanisation of manure removal. This type of group box may be appropriate for an open front calf house, and this combination will be tried out. In our opinion it is possible in this way to save on energy costs. A possible advantage of this system may also be the somewhat easier marketing of the strawy manure in comparison with the liquid slurry. More comprehensive experiments with group housing of veal calves will no doubt answer the question of whether group housing can compete, especially from the point of view of health (which we must take into account for welfare), with the commonly-used individual pens.

REFERENCES

Bogner, H., 1980. Housing and feeding of calves. Animal Regulation Studies 3, 23-28.

Elshof, W.J. and van Putten, G., 1980. De ligbreedte van vleeskalveren in afhankelijkheid van gewicht, borstomvang en staltemperatuur. IVO-rapport B-158, Zeist.

Graf, B., Wegman, R. and Rist, M., 1976. Das Verhalten von Mastkälbern bei verschiedene Haltungsformen. Schweizerische Landwirtschaftliche Monatsheften, 54, 333-355.

Jongebreur, A.A. and Smits, A.C., 1978. Boxafmetingen voor vleeskalveren. IMAG-publikatie 110, Wageningen.

Jongebreur, A.A. and Zwakenberg, A., 1976. Der Einfluss der Boxenabmessungen auf Produktionsmerkmale und Verhalten von Mastkälbern. Züchtungskunde 48: 407-418.
Krippl, J. and Burgstaller, G., 1970. Vergleichende Untersuchungen über den Einfluss unterschiedler Haltungssysteme auf Entwicklung und Schlachtqualität von Mastkälbern. Zuchtungskunde 42: 24-31.
Pesch, W.A., 1970. Drinkautomaten voor kalveren. Denkavit aktualiteiten 32: 4-10.
van Putten, G. and Elshof, W.J., 1974. Der Platz der von einem Mastkalb in Anspruch genommen wird bei 'unbehindertem' Liegen. Rapport C-238, IVO, Zeist.

DISCUSSION

G. van Putten *(Netherlands)* Dr. Jongebreur, could you tell us what was the morbidity rate? You gave figures for mortality rate which was 2.6% in the individual housing and 7.6% in group housing, but what was the morbidity rate?

A.A. Jongebreur *(Netherlands)* Dr. Steenkamer is covering the health aspects of these experiments, so I will leave this question for him to deal with.

J.Y. Blom *(Denmark)* Have you considered whether the mortality rates in the two systems are significantly different? You used so few animals that if you make a simple chi-square comparison between the two groups the figures 2.6 and 7.6% are not significantly different.

A.A. Jongebreur I showed these figures to reveal a tendency that there were more health problems in the group housing system than in the individual housing system. However, I do think it is too early to draw any final conclusions and not worthwhile to talk about the significance of these figures.

W. Goldhorn *(CEC)* With calves kept on straw bedding, how did you measure the straw intake?

A.A. Jongebreur Straw intake was calculated from the supplied straw. The straw provided for the floor which the calves could take up could not be measured. The straw intake was inclusive of straw wastage and we did not measure what amount of the straw was wastage. This also applies to the individual pens.

N. Steenkamer *(Netherlands)* In this preliminary report the intake of straw is given in grammes and this could be misleading. When talking about the results on automatic milk dispensing machines, you made a general statement that this type of production is out of the question for the next five years. I do not think this generalisation should be made since there are some conditions where already now or in the near future these machines can be of use, both in practical and economic terms, in the production of veal.

Firstly, production must only take place with light carcases. The cost of producing the heavy veal calf on the machine is very high. Secondly, as in the example given by Dr. Webster, it can be used for the cheap calf. For example, the Hereford Friesian heifer calf which has no value for other purposes and so can be bought cheap. Even when the production costs are higher the final cost price is still acceptable for the market for they can be killed early at a low weight. Moreover, in some parts of Europe, especially in Switzerland and southern Germany, where there are specific breeds and crossings, machine rearing is feasible if once again the final weight is not too high.

A.A. Jongebreur In my paper I referred to experiments carried out about 10 years ago. Therefore, I did not talk about the feasibility of milk dispensing machines in the future, but why group housing systems were disregarded and why 10 years ago milk dispensing machines could not bring the group housing system back onto the scene.

P.J. O'Connor *(Ireland)* In the individual pens, is there metal sheeting on the side walls at the head-end of the pens?

A.A. Jongebreur There is timber covered by asbestos.

P.J. O'Connor What is the purpose of the asbestos?

A.A. Jongebreur It is used to increase the durability of the timber. It protects the timber from the licking of the calves.

ALTERNATIVE HOUSING SYSTEMS FOR VEAL CALVES, THEIR EFFECT ON WELFARE AND PERFORMANCE AND THEIR ECONOMIC FEASIBILITY

N. Steenkamer
Research Institute for Animal Husbandry 'Schoonoord',
PO Box 501, 3700 AM Zeist, The Netherlands.

INTRODUCTION

Some months ago when I was asked to give the title of the paper to be read here, I suggested:

'Alternative housing systems for veal calves, their effect on welfare and performance and their economic feasibility'.

Afterwards, however, in consultation with Dr. Jongebreur, it was decided that I would concentrate on the health aspects of calves kept in various housing conditions with regard to the same animals and the same experiments on which Dr. Jongebreur reported. In the comparison to be made I would also, however, like to refer to another loose housing system connected with feeding through automatic milk dispensing machines and the whole is placed in the context of the impact on welfare of the animals, the topic of this seminar.

Very often sanitary conditions, in the negative sense of morbidity, or incidence of illness, are undervalued as a factor determining welfare. Ethologists tend to concentrate on behavioural aspects and to point to the state of health as being of secondary importance.

As a result the importance of pain is not given enough recognition. Physical pain is inseparable from disease. Most diseases are painful: arthritis, lung diseases, gastro-enteritis and diseases of the skin. Even ecto-parasites can cause pain.

I do not want to discuss statements to the effect that suffering pain affects animals more deeply, more intensively than men, because of the fact that human beings can talk about their suffering and can escape in religion and culture, thus finding relief: all this has been said before.

Pain must be seen as an attack on animal welfare, and so the connection is made between sickness and welfare.

The comparison to be made is between 3 different housing systems of veal calves:

1. Calves housed in single pens;
2. Calves housed in multiple pens in small numbers (3 - 5) together;
3. Calves in a loose house system where groups of 15 - 50 animals are kept together and fed via an automatic milk dispensing machine.

REARING IN SINGLE PENS

It is a well-known fact that the majority of veal calves have been kept and are still kept in single pens, and this is true of at least 80% of veal calves reared in EC member countries.

These pens,(often condemned as being 'cages' which are too small) if they are made of the right materials and are not worn out (the slatted floors tend to deteriorate first) provide:

1. Good isolation from other animals, which are potential carriers of diseases. They provide a minimum of opportunity for contamination.
2. Very good hygiene - the slatted floor made of hard wood remains clean, free of excrements.
3. Adequate food rationing, which is both good for health and economically sound.
4. Close health supervision, as the way the animals drink gives much information on their condition.
5. Easy observation of the excrements in order to detect possible digestive disorders.
6. A very low incidence of navel sucking and hoof trouble.

From this enumeration it can cautiously be concluded that this type of housing, so much discriminated against, can be of benefit to the young calf. That this applies to young calves must be emphasised, as in the first month of life the young calf is very susceptible to infections (the immune status of these animals is still very low) and as a result of the induced stress of transport, change of food and of environment, the young calf is liable to digestive disorders and respiratory trouble.

The single pen seems to fulfil many of the primary requirements, such as avoiding the spread of diseases in a calf house and easy detection of health problems. Those who have a reasonable dislike of single pens should not ignore these facts, which are supported by abundant evidence.

The inconveniences of the single pen, from both a physiological and a veterinary point of view are:

1. Under stress, calves which are only a few days old have a negative energy balance, and cannot withstand calf house temperatures below 15°C. A slatted floor without straw bedding, even when the wood is dry, does not prevent heat loss and does not contribute to comfort as straw bedding does. Calves in single pens could be given straw bedding, but in practice the costs of labour, in particular, are prohibitive. Part of this negative effect could be compensated by heating the calf house to 15 - 20°C, but fuel costs have increased, so that heating is often not sufficient in practice for the very young calf under 2 weeks of age.
2. Where wooden pens are not properly maintained, broken slats can cause leg injuries.
3. Strangulation is another risk but need not occur in pens which have been properly designed.

REARING IN MULTIPLE PENS

In the multiple pens rearing systems to which I refer, calves are kept on either a wooden or concrete slatted floor covered with straw. Clean, palatable barley straw is offered *ad libitum* from a hay rack or a circular feeder.

Where five calves are kept together the size of the pen is 3.00 m x 2.75 m, providing about 1.65 m^2 per calf. Calves reared on machines are generally stocked more densely: 1.2 - 1.4 m^2 is available per calf.

Bucket fed calves in single and multiple pens receive 2 milk feeds a day in quantities increasing from an initial 1 - 1.5 l/feed when calves are approximately 1 week of age to 8 l/feed in the last month before slaughter. The milk is fed at blood heat in concentrations varying from 100 - 200 g milk replacer/l.

Machine fed calves also receive warm milk which is drunk from rubber teats (1 teat for 15 calves). The milk is offered *ad libitum* at increasing concentrations (from 8% - 22% dry matter).

For several reasons, which have already been dealt with, alternatives to housing in single pens have already been the subject of experimentation for many years. Some experimental set-ups were disappointing, but the multiple pen for 3 to 5 calves, individually fed milk replacer by bucket, and the loose house rearing of larger groups consisting of 15 to 50 calves on automatic milk dispensing machines fed *ad libitum* from rubber teats, may well become acceptable practice in rearing veal calves.

It is still too early, however, to draw far-reaching conclusions or to advocate general application. Not many practical veal producers, on the basis of the evidence produced so far, will risk their existence by converting their calf houses and installing multiple pens for bucket or machine feeding.

Keeping veal calves in groups can be attractive from an ethological point of view (Jongebreur, this publication) but to make a success of group rearing of veal calves it is essential that:

calves in a group are of the same age and vitality;

navel sucking and urine drinking are restricted to a minimum;

the health of the calves is closely monitored;

sick calves are treated efficiently and in time.

In practice it still is difficult, if not impossible, to meet these requirements.

Despite all the enthusiasm for these alternatives these facts should not be ignored or underestimated. A lot of work has yet to be done to find improvements which will make these alternative housing systems applicable in practice and will make them work in large-scale production units.

VETERINARY ASPECTS

Housing conditions to a certain extent determine how veterinary treatment can and will be performed. Table 1 shows

TABLE 1

SOME VETERINARY ASPECTS OF VARIOUS HOUSING SYSTEMS FOR VEAL CALVES

	Single pen	Multiple pen 3 - 5 calves	Multiple pen 15 - 50 calves on automatic milk dispensing machines
Spreading of diseases	x	xx	xxxx
Tracing of sick calves	easy	more difficult	difficult
Examination of patient	easy	more difficult	difficult
Individual treatment:			
1st injection	easy	more difficult	difficult
2nd and further injections	easy	difficult	very difficult (more persons needed and catching device)
Individual oral treatment	possible	difficult	impossible
Dietary measures	easy	difficult	only whole group
Result of diseases	few stragglers	more stragglers	very uneven groups
Standard deviation growth veal calves	10	30	⩾ 30

the differences in this respect between housing calves in single or multiple pens.

This table shows that morbidity is higher with calves in multiple pens. Keeping more calves together, particularly when they drink from the same rubber teat as in the case of machine-rearing, leads to mass contamination and subsequently outbreaks of disease.

The rate of mortality, however, is not necessarily higher. The incidence of certain diseases and infections is influenced by the housing systems in discussion here, as is demonstrated in Table 2.

It is obvious that the major disadvantages of housing calves in multiple pens need to be eliminated before general acceptance of this system can be expected. The problems should encourage ethologists, zootechnicians and veterinarians to continue experimentation using both laboratory and field trials,

TABLE 2

INCIDENCE OF INFECTION/DISEASE IN VARIOUS HOUSING CONDITIONS

	Single pen bucket fed on slatted floor no roughage	Multiple pen 3 - 5 calves bucket fed straw bedding	Multiple pen 15 - 50 calves teat fed from automatic milk dispensing machines *ad libitum* straw bedding
Ectoparasites			
lice	x	xx	xxx
ringworm (*trichophyton verrucosum*)	x	xx	xxxx
Navel sucking/urine drinking	(x)	xxx	x
Hairballs in rumen	xxx	-	-
Coughing	xx	xxx	xxx
Abomasal ulcers	x	xxx	?
Diarrhoea	xx	xx	xx
Hoof trouble	-	-	-

in order to find improvements and solutions for problems not yet solved.

ECONOMIC FEASIBILITY OF HOUSING OF CALVES IN MULTIPLE PENS

In the comparative trials carried out over the last two years, in which calves were housed in multiple pens each holding five animals, economic aspects have also been studied.

In a comparison of calves kept in single pens with those calves kept in multiple pens (n = 78) the following conclusions could be drawn:

1. Performance as expressed in weight gain and feed conversion ratio was similar and slightly better.
2. Carcase quality was the same in both cases. On average, the positive financial result of 1 and 2 was H.fl. 15.00 per calf.
3. Labour costs were higher. The extra labour needed per calf was 140 min (in the period of 150 days) at H.fl. 17.50 per hour; thus the extra labour costs were H.fl.

40.00 per calf. Labour costs could have been lower had the multiple pens not been placed in a converted traditional calf house which was not specifically designed for this type of calf rearing.

4. Costs for straw used for feeding and bedding were H.fl. 27.00 per calf (for 27 kg barley straw for feed at H.fl. 0.20 per kg and 115 kg wheat straw for bedding at H.fl. 0.18 per kg).

Since other costs, such as medicines, were almost the same for both systems, the increased costs of production were H.fl. 40.00 + H.fl. 27.00 - H.fl. 15.00 = H.fl. 52.00 higher per calf housed in the multiple pens. This has to be placed in the context of the net profit per calf over the last five years being less than H.fl. 50.00.

With machine-fed veal calves (data on 400 calves on automatic milk dispensing machines) the economics, in comparison to rearing calves in single pens, were as follows:

Cost price

Lower for lower investment in building and installation	H.fl.	7.50/calf
Higher for less favourable feed conversion/growth	H.fl.	45.00/calf
Higher for use of straw	H.fl.	25.00/calf
Lower for labour costs	H.fl.	7.50/calf
Results in higher production costs amounting to	H.fl.	55.00/calf

However, because of the lack of uniformity of the calves at sale the price paid was, on average, H.fl.20.00 lower per calf. As a result of the higher cost price and the lower selling price, the financial result was a loss of H.fl. 75.00 on the machine-fed calves kept in groups of 15 - 50 calves on straw.

Although these figures on the economics of the two alternative housing systems for veal calves can be challenged, it seems to be indisputable that both health conditions and economics need to be improved considerably before these loose housing systems will be suitable for practical application.

DISCUSSION

A.J. Webster *(UK)* I would like to present two sets of figures to show our performance and health data with the straw yard system. I do agree with just about everything that Dr. Steenkamer has said but I will show what our comparisons have been rearing veal calves in straw yards against those in crates.

TABLE 1

PERFORMANCE OF VEAL CALVES - UNIVERSITY OF BRISTOL

	Crated calves Friesian ♂ bulls	Straw yards Friesian bulls	H x Fr heifers
Finish weight (kg)	205	172	158
Daily liveweight gain (kg)	1.34	1.29	1.17
Carcase weight (kg)	119	98	90
Food conversion ratio	1.56	1.69	1.66
Deaths and culls (%)	12	11	8
Downgraded (%)	None	14	8

Table 1 shows that in straw yards we cannot achieve the same weights although I think with further study these figures could be improved. Our daily liveweight gains are not quite comparable although the difference, in terms also of food conversion ratio, is not as big as Dr. Steenkamer suggests from what he has seen in commercial practice. Our food conversion ratios are again not quite as good but are still good figures. I would add that these calves have nearly all received growth promoters and the veal calf in a straw yard on growth promoter does do very well. The Friesian in a straw yard without growth promoter does not finish. Deaths and culls present no significant difference yet. The percentage of animals downgraded does present a problem. Particularly with Friesians there are a number of animals that have to be sold underfinished, they are too light, and this is pulling down our overall sale income. This is a problem particularly with Friesians and without growth promoters. Using growth promoters we have so far not had any animals downgraded. The Hereford Friesians were mostly downgraded due to respiratory disease.

I would like to make clear that we are comparing what we consider to be two very good systems in their own rights. The housing in both circumstances is good. Some of the British data put out by VOLAC compared a dreadful crate housing system with modest straw yards but without very pronounced veterinary attention. It is easy not to treat an animal in a straw yard if you do not see it is sick. We think we are rather better at recognising disease early.

Table 2 shows that despite the fact that we consider we have good housing and control, we are getting more respiratory disease in groups in straw yards than we do in individual crates. This is a very complex story which I do not want to go into now. However, we are certainly not getting more enteric disease and this again supports what Dr. Steenkamer has said. This is despite the maximum opportunity for contagion with calves in contact with one another. The total overall number of calves treated is really not

TABLE 2

HEALTH SUMMARY - VEAL CALVES AT THE UNIVERSITY OF BRISTOL

	Incidence re 100 calves Crates	Straw yards
Deaths	4	7
Culls	7	3
Calves treated, respiratory	0	24
enteric	30	20
respiratory and enteric	7	8
other	4	11
Origin, from market/dealer	48	77
from farm of origin	52	23

significantly different between the two systems, which is just as well since the greatest proportion of our calves in straw yards were brought in from dealers rather than direct from farm of origin which makes life more difficult for them at the start. So, when well managed, neither of these health records are ideal yet, but they are better than many other British records. What we can say is that with the possible exception of respiratory disease, which has not been too severe a commercial problem, the health of our calves in good straw yards and crates has been no worse, and certainly considerably better than it was in the British type crates which were, I admit, extremely bad by comparison with the best Dutch crate. Therefore, we think that already the straw yard is nearly as good as the best crating system in terms of health.

N. Steenkamer *(Netherlands)* I appreciate what Dr. Webster has said. He has found a higher incidence of respiratory trouble and the same for scour. I must add that these calves have been produced without the help of any anabolic agents which was not said before.

G. van Putten *(Netherlands)* By the term growth promoter, do you mean hormones?

A.J. Webster Yes, oestrogen and progesterone for the males, oestrogen and testosterone for the females.

N. Steenkamer And it has been said that in practice machine feeding has a future but only if efficient growth promoters are used, otherwise there is no future. This is simplifying perhaps but when comparing results, of using or not using a growth promoter, then the truth can be seen.

A.J. Webster I agree that with growth promoter, machine feeding works beautifully.

V. Jensen *(Denmark)* Do you think computerised feeders have a future?

N. Steenkamer We don't know. We have used computerised feeders in rearing calves to wean them gradually off the milk. The results are not bad. On farms where the feed for milking calves is controlled by computer, then the same computer can be used for veal calves.

ETHOLOGICAL AND PHYSIOLOGICAL CONDITIONING OF YOUNG CATTLE IN INTENSIVE HOUSING SYSTEMS

U. Andreae, H.-H. Thielscher, J. Unshelm and D. Smidt
Institute of Animal Husbandry and Animal Behaviour, Mariensee,
Agricultural Research Centre Braunschweig-Völkenrode,
Federal Republic of Germany.

INTRODUCTION

Intensive housing systems for cattle are almost identical to slatted floor housing. In this context the term 'intensive' means a minimal living area per animal and a minimum of work per animal for the farmer. Both of these are provided with slatted floor housing. This housing system is therefore increasingly being used in cattle production.

From practical experience it has been learnt that young cattle should be trained and adapted to slatted floor conditions at an early age. However, it is not known at what age this should occur. With respect to slatted floor constructions for young cattle and fattening bulls, consisting of slats measuring 12 - 15 cm in width with slits measuring 3.5 cm between the slats, an approximate age of 4 to 6 months or a maximum weight between 250 kg to 300 kg (Koller and Süss, 1974) has been recommended for early conditioning of young cattle to slatted floors (Lehmann, 1968; Lippmann, 1971; Bogner, 1981).

Scientific investigations were therefore initiated two years ago at our Institute to obtain basic information on the age at which cattle adapt best.

PRESENT KNOWLEDGE

The following physiological and behavioural parameters for measuring adaptation (Smidt et al., 1980) have been checked:

1. Blood pressure
2. Heart rate
3. Alkaline phosphatase
4. Creatine kinase
5. Lying down intentions per lying down action
6. Lying intervals per day
7. Duration of lying periods per day.

At first 12 male cattle of 9 months of age were observed. They were fitted out with telemetric blood pressure equipment implanted into the carotid artery and kept in place by a girth or strap tied around the animal's thorax (Thielscher et al., 1980). Before this treatment, six of these bulls were housed on deep litter and six other comparable animals were housed on slatted floors from the beginning of the fourth month of life. The reactions of both groups are shown in Table 1.

TABLE 1

COMPARISON OF BLOOD PRESSURE AND HEART RATE IN ADAPTED AND UNADAPTED FATTENING BULLS ON SLATTED FLOOR CONDITIONS

Behavioural situation	Adapted (n = 6)		Unadapted (n = 6)		Difference unadapted/adapted	
Physiological parameters	1st day $\bar{x}$ (s)	2nd day $\bar{x}$ (s)	1st day $\bar{x}$ (s)	2nd day $\bar{x}$ (s)	1st day	2nd day
Standing (80 observ.)						
Syst. mm Hg	99 (20.1)	98 (11.9)	113 (7.6)	96 (9.7)	+ 14**	- 2
Diast. mm Hg	59 (15.9)	60 (11.5)	71 (7.7)	56 (11.2)	+ 12**	- 4*
Heart rate/ min	96 (10.2)	94 (14.1)	99 (9.5)	102 (5.3)	+ 3	+ 8**
Lying (80 obser.)						
Syst. mm Hg	125 (15.6)	127 (9.8)	134 (9.8)	126 (9.5)	+ 9**	- 1
Diast. mm Hg	86 (13.7)	87 (9.8)	98 (8.9)	83 (12.2)	+ 12**	- 4*
Heart rate/ min	85 (10.7)	83 (11.1)	91 (10.0)	94 (6.6)	+ 6**	+ 11**

*P = 0.05 **P = 0.01

Within the first 12 h on slatted flooring, the unadapted bulls showed a relatively pronounced increase of the systolic and diastolic blood pressure as compared to adapted animals. The heart rate in standing or lying bulls, however, rose more during the second than during the first day on slatted floors in contrast to blood pressure which, in unadapted animals, fell below the values of controls. Therefore these preliminary blood pressure measurements seem to indicate a very fast adaptation to slatted floor conditions.

Another methodical approach for testing adaptation ability consists of the measurement of enzymatic activities, especially that of alkaline phosphatase and creatine kinase as indicators of skeleton strain or muscle strain respectively (Unshelm, 1980).

Figure 1 shows that no skeleton strain was found but there was a tendency to muscle strain on the 5th day of housing on slatted floors.

On the 12th day however, creatine kinase values appeared to be normal. In our interpretation, the muscle system of the bulls, probably strained by the insecure walking on slatted floors, adapted itself within 12 days.

An impression of the behavioural parameters of male and female young cattle observed during four 24 h periods from the time of transfer to standardised slatted floors to the 11th day of stay there is given in Figure 2.

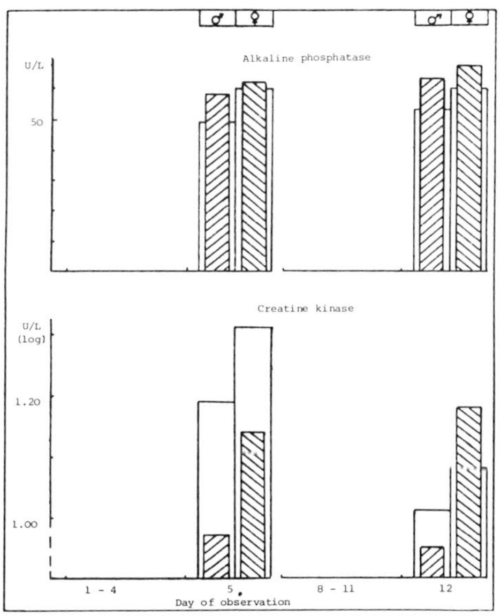

Fig. 1. Comparison of unadapted ☐ and adapted ▨ young cattle (9 months) on slatted floor conditions (n = ♂ 28; ♀ 30)

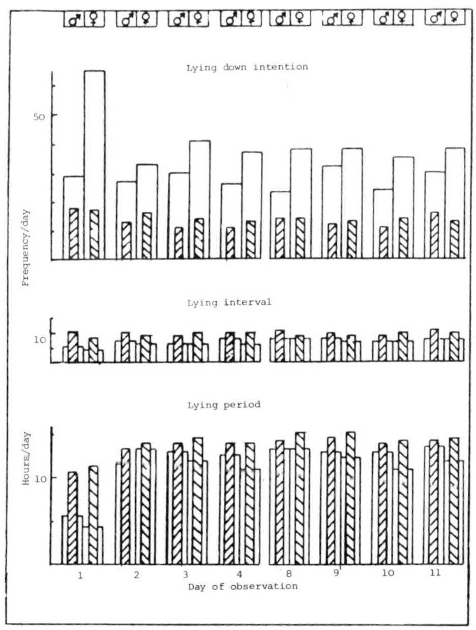

Fig. 2. Comparison of unadapted ☐ and adapted ▨ young cattle (9 months) on slatted floor conditions (n = ♂ 28; ♀ 30)

At present, the frequency of lying down intentions per lying down action, consisting of olfactorial inspections of the ground or the floor, appears to be the most sensitive parameter. In addition, the frequency of lying intervals and the duration of lying periods per day are suitable parameters for completing the ethological characterisation.

In the beginning, as indicated by the behavioural parameters, the 9 month old unadapted young cattle reacted very strongly after being transferred to the slatted floors, and surprisingly, females reacted more than males (Andreae et al., 1980).

During the course of the second observation period, including the time from the 8th to the 11th day on slatted floors, the lying down behaviour of the unadapted animals partially approached that of the adapted ones, but the differences in the three behavioural characteristics remained mostly significant. It must be concluded that, as compared to the controls, young cattle of 9 months of age were not fully adapted to slatted floors within 11 days.

Recent investigations in female cattle of 6 months of age show similar results: the lying down behaviour of unadapted animals is markedly impaired in comparison to the adapted ones up to the 11th day on slatted floors. In addition to this, the control animals which had been kept on slatted flooring for three months were running and jumping on it without hesitation. This playing behaviour of the animals must be defined as normal and demonstrates that complete adaptation to slatted flooring is possible within 3 months (4th to 6th month of age).

With regard to animal welfare, the first question is, how many days should be considered as appropriate for young cattle to adapt to slatted floors? Our present impressions are that cattle should be adapted to slatted floors within two weeks rather than within two months.

The next question is, at what age are young cattle most likely to adapt to slatted floor conditions? Up to now, the frequency of lying down intentions and the time interval between feeding in the morning (7 h) and the moment of first lying down on the first day on slatted floors have been checked in 3 week and 3 month old calves and in young cattle of 6 and 9 months of age, respectively. The data from these observations are presented in Figure 3.

It can be seen from the quantitative differences between lying down intentions that female cattle again reacted more sensibly than male cattle after having entered the slatted floors. Furthermore, the youngest calves seemed to react more intensively than the animals of 3, 6 or 9 months of age. On the other hand, the time interval between feeding and first lying down was the shortest in calves of 3 weeks and the longest in the 9 months old cattle.

In order to obtain more information on the reactions of young cattle of 3 weeks and 3 or 6 months of age after having entered the slatted floors, the lying down interruptions were registered at noon on the first, second and third days. The results of these observations are given in Figure 4.

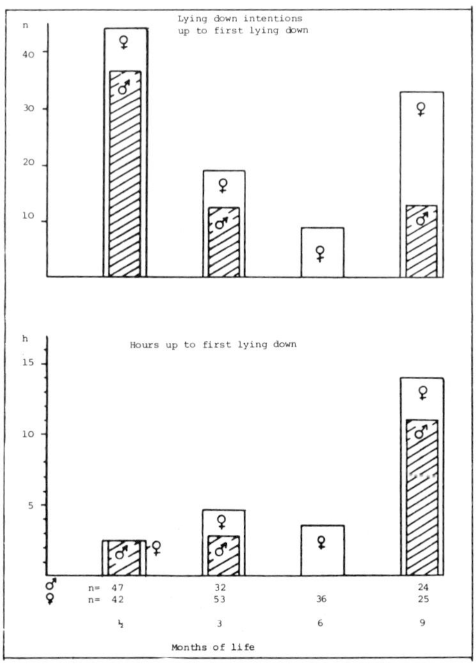

Fig. 3. Lying down reactions in young cattle of different age observed on the first day on a slatted floor.

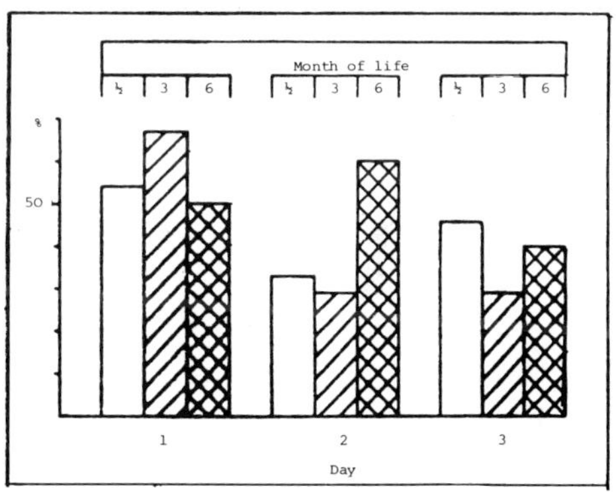

Fig. 4. Number of young cattle of different age (%) showing lying down interruptions on a slatted floor up to first lying down after feeding (1st - 3rd day).

The number of cattle interrupting their lying down actions decreased from the 1st to the 3rd day, which was partly influenced by the age and partly by adaptation effects. But marked differences between the groups were not found. In other words, problems in adapting to slatted floor housing are to be expected in all stages of age.

CONCLUSIONS

The following conclusions with respect to the welfare of young cattle kept on slatted floors might be drawn at present:

1. The adaptation of young cattle to this intensive housing system, especially as measured by the frequency of lying down intentions, can safely be said to have occurred after three months when starting at the 4th month of age, but females show more initial reaction than males.
2. Physiological parameters such as blood pressure, heart rate and creatine kinase give only limited information on the adaptation of young cattle to slatted floor conditions.
3. Calves of 3 weeks or 3 months of age seem to have similar difficulties in adapting to slatted floors as compared to 6 or 9 months old cattle. Therefore, experiments have been initiated aimed at facilitating the first adaptation period by using slats covered with elastic material. During the adaptation period, however, hooves of the animals should be filed down for studying the hoof horn (Schleiter et al., 1973; Brentano et al., 1979).
4. In accordance with pathological hoof evaluations, the ethological recommendation would probably be that slatted floor housing should be started as soon as possible, which means within the first or the second month of life. At this age, which is characterised by a low body weight and correct hooves, there is the best chance that legs and the hoof horn will adapt fully to slatted floors.

REFERENCES

Andreae, U., Unshelm, J. and Smidt, D., 1980. Sind Nutztiere an bestimmte Haltungsverfahren zu gewöhnen? Tierzüchter 32, 184-186.

Bogner, H., 1981. Tierschutzaspekte bei der Haltung von Rindern Sep.-Dr.: Baseler Tagung der Intern. Ges. für Nutztierhaltung, 22/23.1.1981.

Brentano, G., Dämmrich and Unshelm, J., 1979. Untersuchungen über Gelenk- und Klauenveränderungen bei auf Lattenrosten und auf Stroheinstreu gehaltenen Mastkälbern. Berl. Münch. Tierärztl. Wschr. 92, 229-233.

Koller, G. and Süss, M., 1974. Spezielle Haltungsfragen und Tierverhalten beim Rind in Ober J./Koller, G.: Rindviehställe, 2. Aufl., München BLV-Verlag-Ges.

Lehmann, R., 1968. Spaltenboden für Rinderställe. Deutsche Agrartechnik 18, 335-337.

Lippmann, E. and Jeroch, K., 1971. Zur Haltung weiblicher Jungrinder auf Spaltenboden. Wissenschaftliche Zeitschrift der Karl-Marx-Universität, Mathematisch-naturwissenschaftliche. Reihe 20, 506-510.

Schleiter, H., Müller, H.-W. and Spindler, G., 1973. Untersuchungen zur Klauengesundheit bei Kälbern auf Vollspaltenböden Mh. Vet.-Med. 28, 657-662.

Smidt, D., Andreae, U. and Unshelm, J., 1980. 1st 'Wohlbefinden' messbar? Anmerkungen zu einem Tierschutzproblem. Tierzüchter 32, 338-340.

Thielscher, H.-H., Andreae, U., Unshelm, J. and Smidt, D., 1980. Die Blutdrucktelemetrie beim Rind - Ein Beitrag zum Einsatz physiologischer Parameter in der Verhaltensforschung sowie zur Erarbeitung tierschutzrelevanter Indikatoren Dtsch. Tierärztl. Wschr. 87, 214-220.

Unshelm, J., 1980. Verhaltenphysiologische Aspekte der Bullenmast Hohenheimer Arbeiten Heft 107, Tierische Produktion, Grenzen der Leistungsfähigkeit in der Tierproduktion Verlag Eugen Ulmer 1980, 94-110.

DISCUSSION

N. Steenkamer *(Netherlands)* Dr. Andreae stated that calves should be put on slatted floors early in life. Previously, he talked about slatted floors with interspaces of 3.5 cm. Would you recommend using the same slatted floor for the 1 - 2 months old calf and the older animal?

U. Andreae *(FRG)* No, the interspaces should perhaps be 20 mm for the young calves.

H. Eichinger *(FRG)* I think you are going a little too far in saying that heart rate, blood pressure and creatine kinase are not really useful. You put them all together but there is quite a reasonable difference between the reaction of creatine kinase and heart rate for instance. Would you agree with me in this?

U. Andreae This is not a way to assess stress in general, only in this specific problem - adaptation to slatted floors.

H. Eichinger Did you really expect the creatine kinase levels to go up when the animals were put on slatted floors?

U. Andreae Yes, I think that is obvious.

J.Y. Blom *(Denmark)* From an animal welfare aspect, do you consider it is right to put very young calves on slatted floors, even with very narrow interspaces?

U. Andreae More work is needed to study the adaptation period and we will experiment next year on the suitability of using or not using elastic material.

G. van Putten *(Netherlands)* I wonder if you would speculate on what would have happened if the calves had been in cubicles such as adult cattle have. Would adaptation have been easier?

U. Andreae I would hope so. Animals kept on slatted floors from the third to sixth month of life, then going on pasture for three months, had the same adaptation to slatted floors at the age of nine months as animals kept six months on slatted floors.

P.V. Tarrant *(Ireland)* Would Dr. Andreae speculate on what the muscle mechanism of increase in blood CPK might have been? What was the specific event in the muscle that led to the rise in CPK?

U. Andreae From observations, we can see a certain insecurity during the first days of being on slatted floors. This may be connected to muscle strain.

P.V. Tarrant Is there a possibility that muscle damage or bruising may also have occurred when the animals were lying down on a hard or uneven surface?

U. Andreae That is not clear, I do not know.

J.Y. Blom In Denmark, we recommend that calves are not brought onto slats until they weigh at least 125 kg. Most Danish fattening cattle are in fact put on slatted floors but recently quite a few farmers have introduced the Dutch idea of having very narrow slatted floors with cubicles. The farmers have had great success in adapting the calves to the slatted floors using a system where they start the calves at 50 kg in this cubicle slatted floor system and then remove them to the slatted floors at a later stage.

W. Goldhorn *(CEC)* I would strongly support this view from my own experience. I had the impression, Dr. Andreae, that your work was mostly concerned with adaptation of the animal to the surface, i.e. walking ability. This does not take into account the negative heat balance in calves. When farmers buy fattening calves at the age of 5 - 6 weeks, they would like to put them on slats immediately - as you were suggesting, at 1 - 2 months of age. All our farmers have changed and have gone back to using what I referred to previously as a quarantine stable with straw bedding. The calves comfort, albeit subjective, is far greater - they are warmer, they feel more at home than on slatted floors. From practical experience, farmers will now never put calves under 3 months on totally slatted floors. It may be that using cubicles you could combine the two advantages, having warm bedding and the labour-saving of slatted floors.

SUMMARY AND FINAL CONCLUSIONS

F.J. Harte *(Ireland)* I was personally surprised to see the Seminar so orientated towards veal production although admittedly, according to figures produced at this Seminar, the equivalent of about one-quarter of the calves bred in the EC are used for various types of veal production. I would just like to emphasise that the welfare aspects of the 'other' calves also matter. For instance, calves are often poorly treated in transport systems, at markets and, for example, the treatment of internal and external parasites often leaves much to be desired.

We must all realise the importance of veal production. It is a product that is widely demanded by the consumer and it is a biologically efficient way of producing meat though perhaps it is not always economical. Importantly, if we allowed all the present veal calves to grow on into beef, this would put a great excess of meat on the market. We in Ireland have certainly seen in the last 'beef crisis' some of the chaos that can prevail if the market is over supplied with a fresh product, such as meat. So we must be very careful to nurture our veal market and make it as economically and biologically efficient as possible. Obviously, the conditions under which the animals are managed and their general welfare are the most important aspects of this vital meat industry.

I felt throughout this most interesting Seminar that we must all be very careful about what calf welfare really means. Certainly many people have very different views on its meaning, often depending on the environment they themselves live in. I think we all should be very careful not to compare ourselves to animals, and certainly not to ruminants or pre-ruminants and Dr. Bell, among others, emphasised that a veal calf is moving towards a ruminant quite rapidly. 'Opinions' are usually, however, very valuable from experienced people and should not be put aside as 'unscientific'. We must, however, bear in mind that they can be biased, no matter how unintentional that might be.

This is why we must continually look for objective parameters. Yet if we just take the first paper and the last one of this Seminar, we can see the great difficulties of experimentation. We were all, perhaps, unnecessarily critical of the experimental material of the first paper and yet at the end we all appreciated both the difficulties of having good experimental material and the problems of really meaningful parameters. We saw the difficulties that can be encountered by measuring pen size and particularly how the calf lying position is different depending on the prevailing temperatures. I think we were all impressed with the possibilities of measuring renal blood flow. Yet we must appreciate that any new measurement always seems interesting. Measuring lung damage by X-ray seemed to us in our experiments to be a most useful way of measuring respiratory diseases, but now we feel that might not tell us the whole answer. Indeed, I might mention that we should be very careful that the measurements we use are standard. Many people talked about respiratory disease but did not say how they measured it. The term 'crowding disease' for respiratory disease, I felt, was very appropriate. Perhaps it was surprising we had so little reference to respiratory diseases in veal calves. One could come away with the view that it was not a problem. Maybe it is not!

When it came to discussing welfare and behaviour one could detect some difference between people. One speaker felt it was adequate to feed a calf once in the first 18 - 24 h of life, another seemed to feel that this would be unfair to the welfare of the calf. In our own work we suckled calves once daily for 1 - 16 weeks without apparent problems.

Everybody seems to like the idea of straw bedding and, importantly, I think everyone appreciated its problems both from a handling and cost point of view. There seems to be little concern about the fact that veal calves were eating some straw. Dr. Webster indicated one major problem regarding the production of veal on straw using an automatic feeder and that was that the calves just simply did not have a high enough intake after the 8 weeks stage. Maybe there are possibilities here of tethering the calves at the bucket feeding period to avoid inter-sucking. It is also possible that the use of zero milk replacers will reduce feed input costs and thus allow us to slaughter calves at the lighter carcase weight, for we are all well aware of the close relationship between price of calf and slaughter weight. However, dropping cow numbers within the EC will probably have the effect of holding up or increasing calf prices in the future relative to present day values. Indeed, in relation to the automatic machines, one might ask whether we are putting enough effort into design; perhaps more emphasis should be put on the right number of calves per group rather than the number of calves that can be catered for by the machine.

Calf production, and in particular veal production, has the difficulty of marrying together efficiency of calf performance (and here input of costs and returns will be very sensitive), the welfare of the calf and, indeed, the contentment of the operator who is doing the job for quite often calf welfare may depend on the operator's work. The quality of the final product will be the most important factor and, indeed, in any 'readjustment' we must be fully aware of the potential of international competition. If in any way we were to cause veal production to be less economically efficient, it might well be that those who over-stress calf welfare might be first to complain that meat prices were too high relative to world prices. So we must be careful not to impose on our producers any disadvantage compared to potential customers over whose production systems we have no control.

J.P. Signoret *(France)* There is some time left for a final discussion. I believe that the problems concerning calf welfare are probably more difficult than those concerning poultry or pigs that we have previously been dealing with. In veal production, or the production of early-weaned calves, we are attempting to keep ruminants under a non-ruminant diet, to keep animals which are used to being outdoors in intensive conditions. This is a more recent development compared to the more traditional methods of rearing poultry and pigs. In my mind this Seminar has given us ideas, directions and criteria for developing management techniques that will meet the welfare needs of the animals.

N. Steenkamer *(Netherlands)* I would first like to invite the ethologists present to investigate the welfare of certain techniques more or less accepted in practice used in the rearing of calves. I would like to focus on a topic raised here, that of weekend-off feeding. I can understand perfectly well why ethologists concentrated on veal production because this system rears animals in a manner that nature did not intend. However, I would like to invite ethologists to concentrate on several other points. These are more or less current practices - weekend-off feeding, once a day feeding, no roughage up to weaning (in the UK no roughage is fed up to weaning on many farms because it saves labour or makes housing easier) and finally, early weaning in relation to welfare. These are the four points that I would like more attention given to by ethologists.

W. Sybesma *(Netherlands)* I would like to propose research on a new technique which improves meat quality if you subject the carcase after slaughter to electrical stimulation. Perhaps veal calves could then be kept with a higher immunoglobulin content which reduces the incidence of too low haemaglobin levels in calves. This leads therefore to an improvement in calf welfare.

P. Le Neindre *(France)* I would like to explain how I understand welfare. An animal is placed in an environment with given characteristics - climatic, housing, nutrition, social, human. Firstly, what is important is to protect the animal and prevent death. We must take account of physiological factors - digestive, immunologic, metabolic, reproductive capacity. Then there are behavioural factors - comfort, activity, immediate activity, resting etc., and there is also the capacity of later activity, i.e. disease. Is this welfare? It is very dangerous to speak about economics because this is such a very large concept. We have to have an overall picture.

N. Steenkamer I would like to make a few comments on the conclusions which the previous speaker was trying to make. First of all, as to machine rearing, certain trials are being done to investigate the optimum number of calves that can be put together on the machine, the optimum number of teats that can be used, the optimum floor space needed per calf. Much work still needs to be done. As far as urine drinking is concerned, the solution is not yet known. In the last trials performed with some 200 calves, we have tried to apply another housing system. In general, we are trying loose housing with small numbers of calves which are bucket fed, as reported by Dr. Jongebreur, which, in terms of performance, give very acceptable results. To avoid this urine drinking we are tethering the calves during the first 4 - 5 weeks, putting some provisional separation between the calves. They are kept tethered on straw from the beginning, tethered, health inspection and treatment are easy and so all the advantages of the single pen are combined at the age of 4 - 5 weeks. In my opinion, they do not miss lying space at the beginning, they behave quite happily, as far as a non-ethologist can see. So, this system has great promise. Other things have been placed in the pens - lick stones, teats fixed to the wall, water supply *ad libitum*. We think this provisional, temporary tethering may be a good solution to stop urine drinking, but the answer is not totally clear yet.

H. Eichinger *(FRG)* Mr. Chairman, having heard about 20 papers now, I am feeling a little unhappy. The papers have been very interesting and I have learnt a great deal, but it seems to be nearly impossible to bring all these results together in a comprehensive way. I propose that we should agree that certain common rules should be established for research into animal welfare, for instance, to describe the several environments: temperature, humidity, air velocity or gas concentration. These should be known so that comparisons can be made. These observations should be seen from an animal's view, not just general assumptions. It is most important for us to compare different results and then we can make more exact conclusions. We should also be able to compare the animals themselves, to know whether they are more or less productive, to be able to compare growth rate, breed etc. There should be standard values available. Finally, perhaps we could standardise some of the various aspects of research work. For instance, when several indications or measurements are made it is very important from the standpoint of circadian rhythms. We have heard a great deal about calf positions, well would it not be possible to devise five or ten standard positions about which everyone would agree? The same could be done for behaviour patterns. Furthermore, it is important to start to think closer to physiological parameters and in this field there is quite a lot of standardisation work to be performed.

J.P. Signoret *(France)* Welfare is not the only aspect of management. We must define it exactly and agree standards. Welfare might be said to mean health and nutrition; there is no welfare if animals are ill or severely underfed. There is also comfort, the importance of climate, movement, individual and social behavioural needs. We must remember all the welfare parameters and standardise our criteria on health, nutrition, physiological aspects of normal behaviour and the man-animal relationship.

LIST OF PARTICIPANTS

AUSTRALIA

Dr. J.E. Melville
Counsellor (Veterinary Services)
Australian Embassy
51-52 Avenue des Arts
1040 Brussels

BELGIUM

Dr. J.M. Bienfait
Faculté de Médecine Vétérinaire
Rue des Vétérinaires 45
1070 Brussels

Dr. M. Casteels
Ministry of Agriculture
National Institute of Animal Nutrition
Carcase and Meat Division
Scheldeweg 68
B-9231 Melle/Gontrode

Dr. B. Nicks
Faculté de Médecine Vétérinaire
Rue des Vétérinaires 45
1070 Brussels

DENMARK

Dr. H.C. Adler
Royal Veterinary and Agricultural University
Bülowsvej 13
DK-1870 Copenhagen V

Dr. A.T. Fensvig
Danish Meat Research Institute
Maglegaardsvej 2
DK-4000 Roskilde

Dr. M. Jespersen
Danish Research Institute for Poultry Processing
Smørkildevej 8
DK-3400 Hillerød

Dr. N.J. Nielsen
Danish Meat Research Institute
Maglegaardsvej 2
DK-4000 Roskilde

FEDERAL REPUBLIC OF GERMANY

Dr. C. Augustini
Bundesanstalt für Fleischforschung
E-C-Baumannstrasse 20
D-8650 Kulmbach

Dr. K. Fischer
Bundesanstalt für Fleischforschung
E-C-Baumannstrasse 20
D-8650 Kulmbach

Dr. E. Kallweit
Institut für Tierzucht und Tierverhalten
FAL
Mariensee
D-3057 Neustadt 1

Dr. G. von Mickwitz
Freie Universität Berlin
Königsweg 65
D-1000 Berlin 37

Dr. W. Müller
Institut für Tiermedizin und Tierhygiene (460)
Universität Hohenheim
Postfach 700562
D-7000 Stuttgart 70

Dr. D.T. Smidt
Institut für Tierzucht und Tierverhalten
FAL
Mariensee
D-3057 Neustadt 1

FRANCE

Dr. R. Dantzer
Neurobiologie des Comportements
INRA
Université Bordeaux II
146 rue des Saignât
F-33076 Bordeaux Cedex

Dr. P. Le Neindre
INRA
Laboratoire de la Production de Viande
CRZV Theix
F-63110 Beaumont

Dr. J.P. Signoret
INRA Comportement Animal
Nouzilly
F-37380 Monnaie

Dr. C. Valin
INRA
Laboratoire de la Production
de Viande
CRZV Theix
F-63110 Beaumont

GREECE

Dr. J. Papanicolaou
Ministry of Agriculture
Veterinary Services
Athens

IRELAND

Dr. P.J. O'Connor
Deputy Director Veterinary Services
Department of Agriculture
Agriculture House
Kildare Street
Dublin 2

Dr. P.V. Tarrant
Meat Research Department
The Agricultural Institute
Dunsinea
Castleknock
Co. Dublin

ITALY

Dr. Maria T. Auxilia
Istituto Sperimentale per
la Zootecnia
Via Pianezza 115
Turin

Dr. A. Romita
Istituto Sperimentale per
la Zootecnia
Via Panvinio 11
Rome

LUXEMBOURG

Dr. E. Wagner
Administration des Services
Techniques de l'Agriculture
16 route d'Esch
Luxembourg

THE NETHERLANDS

Dr. A.R. Gerrits
Institute for Poultry Research
Spelderholt

Dr. A. Hoogerbrugge
Faculty of Veterinary Medicine
Institute of Animal Husbandry
Yalelaan 17
De Uithof
Utrecht

Mr. L.H. Huisman
Director, Research and Advisory
Institute for Cattle Husbandry
Runderweg 6
8221 RA Lelystad

Dr. J.G. van Logtestign
Department of Science of Food
of Animal Origin
Faculty of Veterinary Medicine
Biltstraat 172
Utrecht

Dr. G. van Putten
Research Institute for Animal
Husbandry 'Schoonoord'
PO Box 501
3700 AM Zeist

Dr. W. Sybesma
Director, Research Institute for
Animal Husbandry 'Schoonoord'
PO Box 501
3700 AM Zeist

UNITED KINGDOM

Mr. A. Cuthbertson
Meat and Livestock Commission
Milton Keynes
Bucks

Dr. T.M. Leach
Agricultural Research Council
Meat Research Institute
Langford
Bristol
BS18 7DY

Dr. D. Lister
Agricultural Research Council
Meat Research Institute
Langford
Bristol
BS18 7DY

Mr. R. Moss
Ministry of Agriculture, Fisheries
and Food
Hook Rise South
Tolworth
Surrey

Dr. C. Platt
World Society for the Protection
of Animals
106 Jermyn Street
London SW1

Dr. D.B. Stephens
Institute of Animal Physiology
Babraham
Cambridge
CB2 4AT

Mr. M.E.T. Watts
Divisional Veterinary Officer
Ministry of Agriculture, Fisheries
and Food
Hook Rise South
Tolworth
Surrey

CEC

J. Connell

W. Goldhorn

J. Wyllie